Global Production and Consumption of Fast Food and Instant Concentrates

Inna Vladimirovna Simakova
Saratov State Vavilov Agrarian University, Russia

Victoria Nikolaevna Strizhevskaya
Saratov State Vavilov Agrarian University, Russia

Roman Lvovich Perkel
Peter the Great St. Petersburg State Polytechnic University, Russia

Galina Yuryevna Rakhmanova
Office of the Federal Service for Supervision of Consumer Rights Protection and Human Welfare in the Saratov Region, Russia

A volume in the Advances in Medical Education, Research, and Ethics (AMERE) Book Series

IGI Global
PUBLISHER of TIMELY KNOWLEDGE

Published in the United States of America by
IGI Global
Medical Information Science Reference (an imprint of IGI Global)
701 E. Chocolate Avenue
Hershey PA, USA 17033
Tel: 717-533-8845
Fax: 717-533-8661
E-mail: cust@igi-global.com
Web site: http://www.igi-global.com

Library of Congress Cataloging-in-Publication Data

Names: Simakova, Inna Vladimirovna, 1977- author.
Title: Global production and consumption of fast food and instant concentrates / Inna Vladimirovna Simakova, Victoria Nikolaevna Strizhevskaya, Roman Lvovich Perkel, Galina Yuryevna Rakhmanova.
Description: Hershey, PA : Engineering Science Reference (an imprint of IGI Global), [2021] | Includes bibliographical references and index. | Summary: "This book offers a comparison of healthy food and fast food, presenting an ecological and hygienic assessment of the impact of fast food on the body in human observations and in vivo experiments, and answers if industrial food can serve as the basis for the nutrition of a modern person"-- Provided by publisher.
Identifiers: LCCN 2020055204 (print) | LCCN 2020055205 (ebook) | ISBN 9781799881971 (hardcover) | ISBN 9781799881988 (paperback) | ISBN 9781799881995 (ebook)
Subjects: LCSH: Nutrition. | Convenience foods--Health aspects. | Processed foods--Health aspects. | Food consumption. | Food supply.
Classification: LCC TX353 .S498 2021 (print) | LCC TX353 (ebook) | DDC 363.8--dc23
LC record available at https://lccn.loc.gov/2020055204
LC ebook record available at https://lccn.loc.gov/2020055205

This book is published in the IGI Global book series Advances in Medical Education, Research, and Ethics (AMERE) (ISSN: 2475-6601; eISSN: 2475-661X)

British Cataloguing in Publication Data
A Cataloguing in Publication record for this book is available from the British Library.

Advances in Medical Education, Research, and Ethics (AMERE) Book Series

ISSN:2475-6601
EISSN:2475-661X

MISSION

Humans are living longer now than ever as a result of advances in the medical field. Having the tools available to train knowledgeable and ethical future generations of doctors and medical researchers is essential to continuing to advance our understanding of the human body and develop new ways of treating and curing sickness and disease.

The **Advances in Medical Education, Research, and Ethics (AMERE)** book series highlights publications pertaining to advancements in pedagogical practice for developing future healthcare professionals, research methods, and advancements in the medical field, as well as moral behavior and practice of healthcare professionals, students, and researchers. Featuring research-based book publications that are highly relevant to the healthcare community, this series is ideally designed for library inclusion at medical universities and research institutions as well as personal use by medical professionals, researchers, and upper-level students entering the field.

COVERAGE

- Clinical Research
- Conflicts of Interest
- Ethics in Medicine
- Healthcare Pedagogy
- Medical Curricula
- Medical Simulation
- Patient Data
- Professional Development
- Research Methods
- Scientific Misconduct

The Advances in Medical Education, Research, and Ethics (AMERE) Book Series (ISSN 2475-6601) is published by IGI Global, 701 E. Chocolate Avenue, Hershey, PA 17033-1240, USA, www.igi-global.com. This series is composed of titles available for purchase individually; each title is edited to be contextually exclusive from any other title within the series. For pricing and ordering information please visit http://www.igi-global.com/book-series/advances-medical-education-research-ethics/132365. Postmaster: Send all address changes to above address.

Titles in this Series

For a list of additional titles in this series, please visit:
http://www.igi-global.com/book-series/advances-medical-education-research-ethics/132365

Advancing Health Education With Telemedicine
Mildred Lopez (Tecnologico de Monterrey Mexico)
Medical Information Science Reference • © 2022 • 377pp • H/C (ISBN: 9781799887836) • US $295.00

Handbook of Research on Updating and Innovating Health Professions Education Post-Pandemic Perspectives
Channing R. Ford (Jacksonville State University, USA) and Kimberly B. Garza (Auburn Universty, USA)
Medical Information Science Reference • © 2022 • 462pp • H/C (ISBN: 9781799876236) • US $395.00

Handbook of Research on Assertiveness, Clarity, and Positivity in Health Literacy
Cristina Vaz de Almeida (Instituto Superior de Ciências Sociais e Políticas, CAPP, University of Lisbon, Portugal) and Susana Ramos (Centro Hospitalar Universitário Lisbon Central, Portugal)
Medical Information Science Reference • © 2022 • 423pp • H/C (ISBN: 9781799888246) • US $395.00

Pedagogies for Pharmacy Curricula
Isabel Vitória Figueiredo (University of Coimbra, Portugal) and Afonso Miguel Cavaco (University of Lisbon, Portugal)
Medical Information Science Reference • © 2021 • 248pp • H/C (ISBN: 9781799844860) • US $265.00

Cases on Instructional Design and Performance Outcomes in Medical Education
Jill Stefaniak (University of Georgia, USA)
Medical Information Science Reference • © 2020 • 330pp • H/C (ISBN: 9781799850922) • US $265.00

For an entire list of titles in this series, please visit:
http://www.igi-global.com/book-series/advances-medical-education-research-ethics/132365

IGI Global
PUBLISHER of TIMELY KNOWLEDGE

701 East Chocolate Avenue, Hershey, PA 17033, USA
Tel: 717-533-8845 x100 • Fax: 717-533-8661
E-Mail: cust@igi-global.com • www.igi-global.com

Table of Contents

Preface

In the context of comprehensive globalization and industrialization in the world, there have been major changes in the market for food and services, which led to an increase in the share of fast food in the diet. The number of fast-food restaurants is increasing annually, and the volume of industrial production of such products is growing rapidly. However, the systematic consumption of fast-food products is a risk of many nutritional diseases associated with impaired lipid metabolism, leading to severe chronic diseases, including cancer. The problem is of particular importance also because the main group of consumers of these products is the younger generation, for whom the negative impact of fat oxidation products on health is especially pronounced.

Over the past 10-15 years, the volume of world production of fast food has increased by about 10 times. In Russia, the annual production of fast-food products reaches more than 200 thousand tons per year and continues to grow rather rapidly.

Among the products of fast-food enterprises, the most popular is deep-frying products. Deep-frying is one of the oldest cooking methods, according to some sources as early as 1600 BC. This method of cooking was widely used by various peoples.

In Chinese cuisine, this method of heat treatment is one of the most popular, and there are several technological methods of frying in a large amount of fat, which differ in the ratio of fat and product and the process of frying: frying in semi-deep fat (chao), frying in semi-deep fat (liu), frying in deep fried.

In ancient Mexico, frying corn with large amounts of fat was used by the Aztecs to produce the equivalent of modern popcorn; traditional tacos were prepared in a similar way.

French cuisine is known for its fries: French fries, deep-fried fish, breaded and uncoated, cauliflower and many others.

A variety of foods can be deep-fried, among which are the sweet dough dishes (curd balls, donuts, crumpets, fruits, and vegetables in batter).

Over the past decades, the consumption of such dishes has significantly increased due to the expansion of their range and the organization of industrial production, especially due to frozen and pre-fried semi-finished products. Such semi-finished products are actively used by fast food enterprises, especially chain ones.

McDonalds, KFC, Burger King are widely known among the fast-food chains. There are also some large enterprises producing foil-wrapped chips and crackers for long-term storage before consumption. However, the main manufacturers of fast-food products are hundreds and thousands of small restaurants and catering establishments.

Deep-fried fast food products are characterized by a very short production cycle (4-6 minutes) and are intended for direct consumption, which makes it extremely difficult both to assess the safety of deep-fat fat at the time of frying semi-finished products in it, and to assess the safety of the fast products themselves. In practice, such an assessment is most often not performed.

The trend towards fast food consumption is manifested not only in the development of fast-food chains. There is another segment of the production of products intended for consumption at work or at home and requiring minimal effort to prepare. These include soups, main courses such as instant noodles and potatoes, and some sweet foods and pastries. A special place is occupied by instant food products, designed to minimize labor and energy costs.

All these products refer to food concentrates and are now widely used in all countries of the world as products of mass use and widespread consumption by the population at home, as well as in children and dietary food.

Food concentrates currently produced can consist of one type of raw material (mono-concentrates) or, most often, include a mixture of several products selected according to the approved product recipe (complex concentrates). The following groups belong to food concentrate products:

- food concentrates of soups (broths, soups) and second course;
- food concentrates for sweet dishes;
- semi-finished flour products (muffins, mixtures for making pancakes and pancakes, etc.);
- breakfast cereals (cereals, puffed grains, corn sticks, oat diet foods, dry foods for children and diet food);
- instant food concentrates (not requiring cooking).

The range of food concentrate products is currently quite diverse; the popularity of these products is quite high.

Unlike fast food products, food concentrates are intended for subsequent consumption; during their production there is the necessary time for careful control of the quality and safety of finished products, including during the subsequent storage of these products in the retail network.

Deep-fried foods are in demand in both developed and developing countries for their distinct taste, aroma and crunchy texture. At the same time, facts are accumulating indicating that the consumption of deep-fried foods is unsafe for consumers, and the negative effect is associated with the accumulation of thermo-oxidative fat destruction products in finished products.

Further development of the production of deep-fried products is possible only with the unconditional provision of the safety of these products.

In this regard, questions arise: can fast food products and food concentrates be a worthy substitute for a full-fledged rational diet? How safe is regular consumption of these products for human health?

This monograph is devoted to these questions.

Roman Lvovich Perkel
Peter the Great St. Petersburg Polytechnic University, Russia

Inna Vladimirovna Simakova
Saratov State Vavilov Agrarian University, Russia

Introduction

This book is devoted to the comparison of healthy food and fast food, includes an ecological and hygienic assessment of the impact of fast food on the body in human observations and in vivo experiments. The authors of the book raise topical issues of the relationship between nutrition and health, as well as the search for solutions: can industrial food be the basis for the nutrition of a modern person in general, as well as the nutrition of children and adolescents?

The subject area of this work is the technological and medical aspects of the production and consumption of fast food.

This book is the result of 15 years of scientific work of the authors.

The book contains eight chapters, four of which are based on the authors' own research.

Over the past century, there have been significant changes in food technology due to the processes of globalization.

The book analyzes and substantiates the aspects of the relationship between the transformation of food culture, and the possible impact of this process on human health and generation as a whole. The main features of food globalization and development trends are noted, including the pros and cons of globalization, urbanization and industrialization of food.

Two chapters are devoted to the nutrition of a modern person and the development of pathological conditions. Particular attention is paid to the state of children health in the light of malnutrition from birth to puberty and predicting the quality of their health in the future.

The author's classification of fast food products is proposed. The analysis of fast food consumption and leading global companies producing such products is carried out.

Based on the example of the authors' studies, the safety indicators of the most popular fast food products with a high fat content, as well as some industrial fast food products are substantiated.

Animal studies have established the level of toxic effects and pathologies after consumption of fast food products containing various concentrations of secondary oxidation products. Oxidation products cause serious changes in the biochemical composition and blood cells, lead to disruptions at the tissue and cellular levels in the work of the gastrointestinal tract, cardiovascular and excretory systems. Food concentrates have a similar effect on the body, which is associated with both a change in the fat component and the content in them of a large amount of related substances.

The book is intended and will be of interest to the scientific community, graduate and postgraduate students in medical and nutritional specialties, as well as of advanced food companies.

Chapter 1
Transformation of Food Culture in Different Countries:
Globalization, Industrialization, Urbanization

ABSTRACT

In this chapter, the authors considered the influence of the processes of globalization, urbanization, and industrialization on the food culture. The analysis made it possible to show the relationship between the transformation of food culture and the lack of essential components of nutrition. The chapter identifies stable trends influencing the worldview of a modern person about food culture as a whole – the mutual influence of the West and the East. There is a shift from national authentic food systems towards international ones. This trend has driven the urbanized person towards chronic snacking and fast-food culture. The authors considered the risks of the impact of global food technologies on the transformation of the health of generations in the light of the scientifically grounded principles of nutritional science.

INTRODUCTION OF THE CHAPTER

Today, all changes in the world are happening due to the globalization processes (Ilyin I.V., Leonova O.G. 2016). They touched on such an important aspect of human life as nutrition.

DOI: 10.4018/978-1-7998-8197-1.ch001

The starting points for worldwide globalization of nutrition were the media revolution and the migration of people from rural areas to cities, from country to country. The availability of scientific and technical information and the speed of dissemination have a tremendous impact on the transformation of the food culture of different peoples. Traditional approaches to national cuisines are leveled. Total industrialization leads to the spread of adapted technologies everywhere.

Food culture as a whole is centuries-old eating habits, cooking technologies, ethnic traditions of consumption, due to the way of life and lifestyle. Traditions and food culture have their own prerequisites, which are dictated by the formation of adaptive digestion of people depending on living conditions, local habitat, food sources and climate. Depending on the type of food, the enzyme digestion system, intestinal microflora gradually formed, which together with other environmental factors forms the health level.

To ensure and maintain health is not enough to know the basics of proper nutrition. It is the food culture that largely determines the state of human health for many years. A person should know how to eat properly, should have table manners, and observe a diet. An important factor is the observance of moderation in eating.

Food culture provides for the mandatory implementation of food hygiene rules and this is a much larger concept than compliance with hygiene rules. Food hygiene in the concept of nutrition culture is food that is safe for human health, replenishing the body's needs for all necessary substances.

Thus, at the moment, food culture is a high level of knowledge about food consumption, practical skills in product selection, as well as in determination of nutrient demands, depending on the health level and life quality.

The ability to take advantage of the accumulated knowledge that makes up the concept of the nutrition culture of a modern person is expressed in the personification of diets, taking into account the balance of basic nutrients. The main aspect is the physiological state of the body (primarily nutritional status), working conditions, gender and age of the person, climate patterns of the region. According to the results of numerous examinations in different regions of the world, a person in a global society has a deformation of nutritional status associated with an imbalance of essential substances, causing various pathological conditions.

Over the past 20 - 25 years, the view on food consumption has radically transformed. A century ago, the lack of technology that creates the convenience of everyday life of a modern person led to the need for cooking immediately before a meal. This significantly regulated the diet and the amount of intake food.

The pace of life, social and economic status of the population of megacities force them to eat out. The ritual of eating that has developed over the centuries ceases does not have significant significance due to a change in the traditional culture of food consumption. The modern pace of life is the reason of quick bite and development of fast food culture.

Changes in food culture, and as a result of nutritional and digestive status are amplified with the increasing influence of globalization and urbanization.

MAIN FOCUS OF THE CHAPTER

The Main Features of Nutrition Globalization and Development Trends

The main features of globalization appear in the food culture (Vigel N.L., 2016):

- the formation of the homogeneity of the sociocultural space of nutrition - everyone eats the same and monotonous, regardless of nationality, age, family status, etc. This factor is a serious prerequisite for "hidden" hunger - the development of the insufficiency of essential substances;
- universalism - the use of universal technological methods for cooking food with an extended shelf life, the versatility of food ingredients and food additives, the use of universally refined food sources (simple purified proteins, refined fats, purified mono- and disaccharides, etc.). This leads to the rapid digestion of food, and, as a result, the need for a new portion of food;
- simultaneity - the coincidence of food intake with time, movement, i.e. complete violation of the diet, leading to poor digestion. The habit of eating on the move led to the catastrophic popularization of fast food. In addition to the characteristic features of nutrition globalization, one can observe steady trends that affect the modern worldview of a person about the nutrition culture as a whole.

Two trends prevail: Americanization and (or) Westernization of nutrition; Easternisation of nutrition.

Americanization and Westernization of food is associated not only with fast food restaurants, but with technologies for the cooking, storage, transportation and sale of products developed in the West.

The Western economic system of free market and entrepreneurship has turned cooking from homework into industry.

All over the world, preference is given to semi-finished products (products prepared for quick cooking - maximally prepared for heat treatment, or maximally prepared for consumption). This dramatically saves time, improves efficiency, and standardizes product parameters.

Historically, mankind has been looking for a solution to the safety of food products and the extension of their shelf life. Many traditional foods originates from storage technologies: for example, cheese is a way of storing milk that came from Bedouins in the Middle East; beer is a way to preserve water in hot countries (in Sumer and Egypt); and sausages and gourmet foods are a way of meat storage.

Often Americanization and Westernization, being a process of nutrition globalization, are associated with the growing influence of McDonald's and similar franchises. At the same time, the reality is that Western technologies of processing, storage and transportation of products (mass conservation; freezing; packaging and storage technologies in gas media) contribute much more to the establishment of a single global system of food consumption (Andrew Coe, Chop Suey 2009, VeselovYu.V., JinJunkai, 2016).

Easternization of food is associated with the migration of oriental cuisines to Europe and America. There are several types of cuisine there:

- Chinese cuisine - bringing Wok technology and soy sauce to the world;
- Indian cuisine, that is characterized by a large variety of spices and popular "curried chicken";
- Japanese cuisine, that is popular with rolls and sushi;
- cuisines of the Middle East (Arabic, Turkish) with doner kebab, shawarma and others.

Americanization and Easternization are two powerful migration flows in the food culture that influence changes in the food style in the world.

It should also be noted the influence of Italian and Mexican cuisine in modern fast food, that are popular with pizza, burritto and other dishes.

It is important to note that there is a unifying component of the majority of mentioned trends. It is a deep-frying technology, as well as the prevailing starch content in fast food products.

With all of the above features and trends in the globalization processes, the nutrition influence on human life has strengthened with the increased stress factor of modern urbanized life.

PROS AND CONS OF FOOD GLOBALIZATION, URBANIZATION, AND INDUSTRIALIZATION

In terms of globalization, the current level of food production development is characterized by a number of positive effects of economic importance:

- liberalization of foreign economic relations, that simplified the food production processes all over the world. It promotes the development of a competitive environment and equal access to commodity markets. According to the liberalization system, import duties for products from developing and emerging countries can be reduced by up to 50%, and for least developed countries these duties are determined in the annual messages of the UN, up to the complete cancellation. This can significantly alleviate the impact of the global food problem;
- transnationalization of capital and food production is determined by the movement of capital from countries with a relatively excessive amount to countries with a deficit, but characterized by an excess of resource production factors (labor, land, etc.). It leads to the formation and strengthening of the role of multinational food production companies with uniform requirements for quality and safety, predicted characteristics and functional and technological properties;
- regional economic integration associated with the creation of single economic complexes for food production. It is caused by the development of holdings and corporations producing primary products (grain, vegetables and fruits, raising livestock with a full cycle of processing to a finished food product, justified according to the scientific and technical developments of resource-saving integrated technologies;
- internationalization of economic life, that is characterized by technological development methods that contribute to the adaptation of food products to the characteristics of regional consumption. Food products are becoming adapted for potential use in any region, and also for consumers of various ethnic groups;

- unification of the rules of economic life, the creation of a system of interstate regulation of world economic relations in the food production due to unified approaches to the production and processing of food products.

These positive economic features give rise to a number of global problems of modern civilization.

The global food culture is perceived as "freedom from tradition". Modern man is becoming increasingly dependent on high-tech food production, automation, high-tech mass consumption. The fast food monoculture has formed not only new sociocultural stereotypes, but also generates global risks associated with the consumption of alternative types of food (Vigel N.L. 2016).

Technology, urbanization, and change in income are factors that have fundamentally changed the global food system and thereby food consumption (Popkin B.M., Adair L.S. 2012; Zobel E. H. et al 2016).

Intensive urbanization is determined by the growing political and socio-economic role of cities and the increasing migration of the population from rural to urban area.

The population of mega and big cities is growing steadily and continuously. According to the UN, from 1950 to 2014, the share of urban population in the world has grown from 30 to 54%, and by 2050 it will reach 66%. Such rates cause an increase in tension in the urbanized space due to a change in the food culture (there is a growing number of working women who do not have enough time for daily cooking). This, undoubtedly, has already led to the evolution of urban food standards, and a sharply growing demand for prepared foods. The problem caused by nutrition urbanization is a sharp change in nutrition, for example, when changing a place of residence or other factors, which can adversely affect human health.

The growing concentration of the population in large cities causes an aggravation of the problem of uninterrupted supply of urbanized areas with foodstuffs.

That is why one of the problems of globalization is the food problem. In terms of food security, the following types of countries are distinguished (Kovalev E., 2005):

1. The largest developed countries, mainly exporting food in the world (USA, Canada, Australia, some European countries).
2. Small countries actively exporting food products (Finland, Hungary, Belgium, the Netherlands and others).

3. Countries that are plagued by a lack of food, but able to purchase them (South Korea, Japan), these are the main importersю
4. Countries that barely meet their food needs from their own production (India, China, South America).
5. Countries that are plagued by a lack of food, developing water, land, scientific and technical resources to ensure self-sufficiency (Middle East, countries of Southeast Asia).
6. Countries where food security is at a level acceptable by medical standards (countries of the former USSR).
7. Countries with steadily worsening food security and an emerging food crisis (African and Asian countries).

The food problem is currently considered in two ways:

- in a proper or local meaning: food shortages due to insufficient development of productive forces in the agricultural sector of the economy or its individual links, due to adverse weather conditions or socio-economic events in a particular country or region;
- in a broad meaning: in the form of a complex of economic, technical, socio-demographic, political problems of production, distribution, exchange, consumption of food, the lack of which leads to malnutrition, hunger and epidemics, socio-political instability, revolution, rebellion, etc. (Lysak M.A., 2013).

It should be noted that during the last three decades, the production of food products did not just keep up with the population growth, it increased even faster than the population of the planet.

If the food produced worldwide was distributed evenly among all the inhabitants of the planet, then each of them would have a daily portion of food with an energy value of about 10.460 kJ. And this is 628 kJ more than the average norm prescribed for a healthy diet. For example, the most densely populated country in the world, the People's Republic of China, despite the population growth, managed to not only defeat hunger, but even start exporting food.

If we consider the cause of malnutrition and hunger in the world, this is not the absence of natural prerequisites for food production and not too rapid population growth. It is poverty, that hinders the adequate food supply of a significant part of the world's population. Therefore, the starving regions coincide geographically with the regions of poverty incidence.

According to the calculations of British experts, productive forces reached a level of development, at which more than 10 billion people could be provided with food. Considering that almost all the crop goes to the country whose capital was invested in this agricultural enterprise, the volumes of food products are distributed between countries extremely unevenly. The involvement of developing countries in the world economy, their specialization in the production of tropical and industrial crops reduce their food supply. The underproduction of food in developing countries is accompanied by the overproduction of tropical crops in them, which reduces their cost.

Thus, in the vast areas that make up the periphery of the world economy, there is still a lack of funds for the development of agriculture in these countries, and the dominance of outdated production relations in the agricultural sector often makes it impossible to give impetus to the development of agricultural production and the limited obtained resources (Khisamutdinov I.A., 2008).

Hunger and poverty dominate mainly in the states of Africa located to the south of the Sahara desert, in some regions of East Asia and Latin America, as well as in most countries of Southeast Asia (Ziyavitdinova N. M., 2016).

The true extent and severity of the food problem can be estimated on the basis of research provided by the Food and Agriculture Organization of the United Nations (FAO) (**Guijt, J, P**. et al. 2021).

Based on FAO statistics, the number of hungry people on the planet is approximately 500 million people, while about 240 million people are doomed to for illness and death because of starvation. According to the FAO, in 2010-2012, almost 870 million people were chronically malnourished, which makes up 12.5% of the world's population, that is, one in eight people. The vast majority of them (852 million) live in developing countries where 14.9% of the population are malnourished (Table 1 and 2).

Table 1. Undernourishment around the world, 1990–92 to 2014–16

	Number of undernourished (millions) and prevalence (%) of undernourishment									
	1990–92		2000–02		2005–07		2010–12		2014–16*	
	Number	%	Number	%	Number	%	Number	%	Number	%
WORLD	1 010.6	18.6	929.6	14.9	942.3	14.3	820.7	11.8	794.6	10.9
Developed Regions	20.0	<5.0	21.2	<5.0	15.4	<5.0	15.7	<5.0	14.7	<5.0
Developing Regions	990.7	23.3	908.4	18.2	926.9	17.3	805.0	14.1	779.9	12.9
Africa	181.7	27.6	210.2	25.4	213.0	22.7	218.5	20.7	232.5	20.0
Northern Africa	6.0	<5.0	6.6	<5.0	7.0	<5.0	5.1	<5.0	4.3	<5.0
Sub-Saharan Africa	175.7	33.2	203.6	30.0	206.0	26.5	205.7	24.1	220.0	23.2
Eastern Africa	103.9	47.2	121.6	43.1	122.5	37.8	118.7	33.7	124.2	31.5
Middle Africa	24.2	33.5	42.4	44.2	47.7	43.0	53.0	41.5	58.9	41.3
Southern Africa	3.1	7.2	3.7	7.1	3.5	6.2	3.6	6.1	3.2	5.2
Western Africa	44.6	24.2	35.9	15.0	32.3	11.8	30.4	9.7	33.7	9.6
Asia	741.9	23.6	636.5	17.6	665.5	17.3	546.9	13.5	511.7	12.1
Caucasus and Central Asia	9.6	14.1	10.9	15.3	8.4	11.3	7.1	8.9	5.8	7.0
Eastern Asia	295.4	23.2	221.7	16.0	217.6	15.2	174.7	11.8	145.1	9.6
South-Eastern Asia	137.5	30.6	117.6	22.3	103.2	18.3	72.5	12.1	60.5	9.6
Southern Asia	291.2	23.9	272.3	18.5	319.1	20.1	274.2	16.1	281.4	15.7
Western Asia	8.2	6.4	14.0	8.6	17.2	9.3	18.4	8.8	18.9	8.4
Latin America and the Caribbean	66.1	14.7	60.4	11.4	47.1	8.4	38.3	6.4	34.3	5.5
Caribbean	8.1	27.0	8.2	24.4	8.3	23.5	7.3	19.8	7.5	19.8
Latin America	58.0	13.9	52.1	10.5	38.8	7.3	31.0	5.5	26.8	<5.0
Central America	12.6	10.7	11.8	8.3	11.6	7.6	11.3	6.9	11.4	6.6
South America	45.4	15.1	40.3	11.4	27.2	7.2	ns	<5.0	ns	<5.0
Oceania	1.0	15.7	1.3	16.5	1.3	15.4	1.3	13.5	1.4	14.2

**Source: FAO. Data for 2014–16 refer to provisional estimates. T h e s t a t e o f f o o d i n s e c u r i t y i n t h e w o r l d 2 0 1 5 9*

Table 2. The changing distribution of hunger in the world: numbers and shares of undernourished people by region, 1990–92 and 2014–16

	Number, (millions)		Regional share, (%)	
	1990–92	2014–16	1990–92	2014–16
Developed regions	20	15	2.0	1.8
Southern Asia	291	281	28.8	35.4
Sub-Saharan Africa	176	220	17.4	27.7
Eastern Asia	295	145	29.2	18.3
South-Eastern Asia	138	61	13.6	7.6
Latin America and the Caribbean	66	34	6.5	4.3
Western Asia	8	19	0.8	2.4
Northern Africa	6	4	0.6	0.5
Caucasus and Central Asia	10	6	0.9	0.7
Oceania	1	1	0.1	0.2
Total	1 011	795	100	100

**Note: The areas of the pie charts are proportional to the total number of undernourished in each period. Data for 2014–16 refer to provisional estimates. All figures are rounded.Source: FAO.*

It can be seen from the Table 1, that the growth rate of the malnourished population is decreasing, mainly due to the improvement of the food situation in the developing countries of Asia, but nevertheless, the positive growth of the analyzed indicator is explained by the constant increase in the number of malnourished people in the countries of Africa where thy 22.9% of all starving people live.

In contrast, in rich world regions, people often handle food very wastefully, which is another cause of hunger on Earth.

The second most important problem is the problem of hidden hunger. It has become common and relevant for both poor and rich countries in the modern, technologically advanced world. Hidden hunger is a chronic deficiency of vitamins and minerals and other essential substances in the diet. Life in megacities and especially in environmentally disadvantaged cities, constant nervous overload requires increased consumption of vitamins and other substances. The severity of the situation becomes even more pronounced in cold climates, economic poverty, and the lack of certain elements in soil and water. Pregnant women, young mothers, and children are especially affected by deficits.

In addition, modern food production technologies do not contribute to the preservation of valuable essential substances in products due to the intensification of production and processing.

In most countries, emergency measures have already been taken or are being taken to combat hidden hunger. The government clearly recognizes that these measures may prevent a national catastrophe. Mandatory enrichment of mass products has already been embodied in different national programs, enshrined in laws. In the USA, Canada and other countries, it is prescribed to enrich absolutely all flour. In Venezuela, the incidence of anemia in children 7–15 years old decreased from 37 to 19% during one year, while enrichment of school breakfasts in Peru reduced the number of anemic children by almost 5 times in six months. Prevention of micronutrient deficiency in America prevents 4 out of 10 childhood deaths and reduces maternal mortality by a third; it increases the working capacity by 10%, the average coefficient of intellectual development (IQ) by 10-15 points, and the country's gross product by 5% (Maksakovsky V. P., 2011). Such preventive measures can help to overcome the problem of hidden hunger, however, maintaining the natural components of food in their native state is more preferable to ensure the health of generations.

The food problem is global, not only because it affects the peoples of the whole world, but also for a range of another reasons:

- the intensification and industrialization of agricultural production, the creation of modern conditions for the storage and processing of its products are impossible without support from other sectors of the economy and the state. At the same time, the biological value of the manufactured products cannot always make up a deficiency in nutrients.
- orientation to food imports often negatively affects the food production within the country. In addition, most imported products can significantly affect the adaptive nutrition type that has developed over the centuries, and therefore it can provoke the manifestation of latent nutrition-dependent diseases.

It should be noted that the growth of migration leads to a global conflict in nutrition on the background of "overload" due to the demographic aspects of urban life - a large number of people, high population density, its heterogeneity. Highlighted aspects primarily affect food safety. Population flows pose a rather strong risk of migration of pathogenic and conditionally pathogenic microflora in places of public food consumption. In addition to the need to extend the life cycle of food raw materials, semi-finished products and finished products and the production of cost-effective food products, this problem causes the risks of the impact of global food technologies on the transformation of the health of generations.

Nutrition can be considered normal only when food fully meets the needs of the individual organism, ensures constant body weight and contributes to the normal functioning of all organs and systems of the body.

That is, in accordance with modern data based on science, nutrition should provide the following criteria:

1. Nutrition supports the molecular composition of the body and provides its energy and anabolic needs.
2. In metabolic and trophic relations, a man is not an organism, but a supraorganismal system, which include macroorganism, the microflora of his gastrointestinal tract – endoecology. Positive symbiont relationships are maintained between the host organism and the microflora of the digestive apparatus (Ugolev A.M., 1987, 1991).
3. Nutrition and digestion are connected not only with one flow of nutrients released as a result of digestion, but also with at least three more flows. The first is the vital flow of regulatory substances - hormones and hormone-like compounds. This flow consists of two ones: endogenous and exogenous. The first flow includes hormones produced by the endocrine cells of the digestive apparatus, the second one includes the so-called exohormones, which are formed mainly during the enzymolysis in the gastrointestinal tract. The second flow consists of food ballast substances, modified by the bacterial flora of the intestine. It is also biologically important, since secondary nutrients enter the internal milieu. The third flow is the flow of toxic compounds formed from toxic food substances, as well as toxic bacterial metabolites formed in the gastrointestinal tract due to the activity of the bacterial flora. Apparently, this flow is normally physiological.
4. Ballast substances, or dietary fiber, are not ballast, but an evolutionarily important food component. The flow of such substances modified by the microflora of the gastrointestinal tract is necessary for the normal functioning of the digestive apparatus and the body as a whole.
5. The balance of nutrients in the body is achieved as a result of the release of end products capable of absorption due to cavity and membrane (in some cases intracellular) digestion, as well as due to the synthesis of new compounds, including essential ones, by the bacterial flora of the

intestine. The relative role of primary and secondary nutrients varies widely.

6. The role of nutrition in the formation of physiological and psychological standards of a person increases even more as a result of the discovery of the functions of certain amino acids as neurotransmitters and as their predecessors.

Thus, at present, nutrition cannot be interpreted as simply supplying the body with a certain set of chemical elements. This is a complex process in which the gastrointestinal tract interacts with other organs and systems of the body and is a source of a great number of hormonal signals. Endoecology must also be considered. It is represented by a peculiar set of closely interacting bacteria that realize a lot of important transformations concerning both endogenous and exogenous substances. Additional nutrients form in the result of transformational changes in these substances and ballast dietary fiber. It is equally important that the bacterial population of the gastrointestinal tract implements a special type of homeostasis - trophostasis (from the Greek trophos - food, nutrition), that is, maintaining the constancy of the trophic flow from the digestive tract to the internal milieu. In the absence of bacterial flora, our trophic resistance is sharply violated. Normal endoecology can be disrupted by various influences, which causes an increase in the flow of bacterial metabolites and provokes a number of serious diseases (Baranovsky A. Yu., 2012)

Can modern global food guarantee the health of nations? This is the most pressing issue of the present.

This issue concerns innovative food production. Basically, this is the optimization of formulations or a technological process, the purpose of which is the appearance of functional properties in traditional food products. Accordingly, there are products that make up the need for vitamins, or improve digestion, or are lightweight in calories, etc. Moreover, all these products have prolonged shelf life, and it contradicts their declared functionality.

Currently, the following processes affect the development of nutritional-dependent diseases:

- decrease in the general nutrition culture of the population;
- replacement of nutritious food with fast food, due to the intensity of life in megacities;

- the popularity of fast food products produced in large quantities by the food industry (snacks, food concentrates, etc.), popularized by advertising;
- presence of food additives that enhance the impression of food;
- deficiency in nutrients due to the reduced nutritional value of fast food, compared with high-grade diets;
- consumption of foods with hydrogenated fats, defective proteins, simple carbohydrates, etc.;
- the imperfection of the legal framework in ensuring food safety, despite significant changes being undertaken at the interstate and state levels, recommendations of the FAO / WHO.

Particular attention should be paid to risk factors (Figure 1), which have the greatest impact on the health status of the population.

Figure 1. Risk factors, which have the greatest impact on the health status of the population

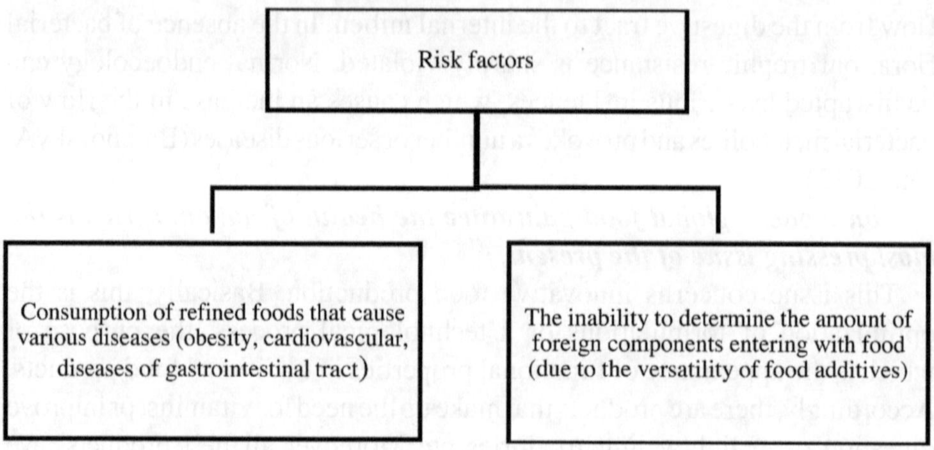

Analyzing Figure 1, it is important to note that most refined foods are foods designed for fast food.

Compared to traditional foods, refined and ultra-processed foods are more energy dense; have more added simple sugars (six times higher than products from groups 1 and 2), sodium and saturated fats and have a low content of dietary fiber (half compared with products from groups 1 and 2) (Monteiro C.A. et al., 2011; Moubarac J.C. et al., 2013).

In Canada, diets dominated by ultra-processed foods exceed the World Health Organization recommendations for fat, sugar, and sodium (Moubarac J.C. et al., 2013).

Data from three promising cohorts, including 120.877 American women and men, shows that weight gain over a 4-year period is associated with a predominance of processed foods (e.g. chips, French fries, sugar drinks, pastries, and meat products) (Mozaffarian D. et al., 2011).

Simple sugar and refined starch predominate in modern foods, increasing the amount of these substances in the human diet. During digestion, the starch excess runs to fat, and since energy consumption is often reduced in megacities, it is stored under the skin as a reserve.

Obviously, global economic growth, an increase in the availability of per capita food and ultra-processed foods as the dominant source of energy have caused global changes in the food system.

The global shift in food supply is seen as the one of the main reasons of the obesity epidemic and the increasing prevalence of concomitant complications such as type 2 diabetes, cardiovascular disease and cancer.

Increased demand and regular consumption of fast food, ultra-processed foods create a mechanism that often leads to the processes discussed in Figures 2 and 3.

Figure 2. Processes that influence over changes with regular consumption of fast food

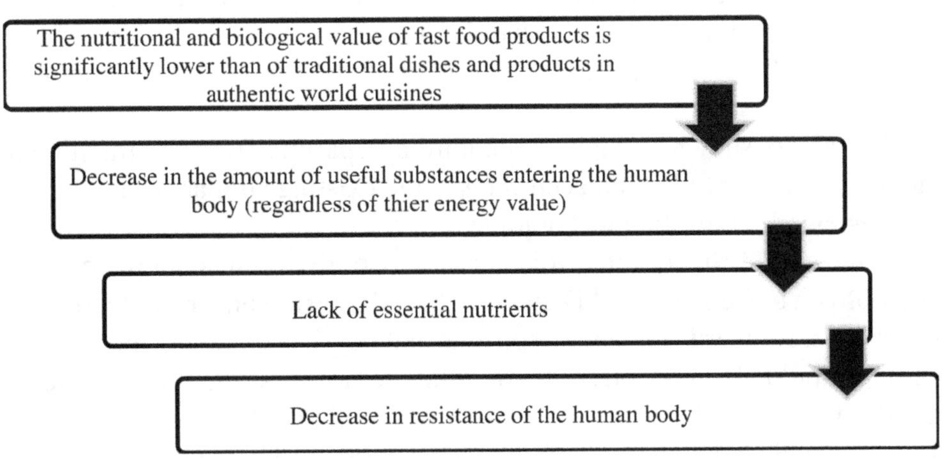

Figure 3. Versatility factors of nutritional supplements

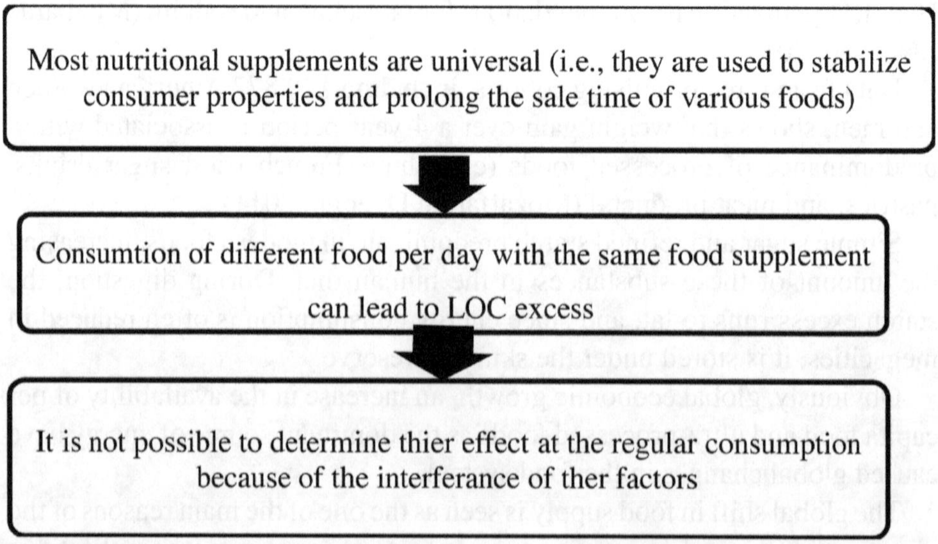

Based on the fact that fast food products have significantly low nutritional and biological value, it is possible to predict the deficiency in useful substances. These criteria are not connected with the compensation for caloric content. In turn, a decrease in t useful substances causes a shortage of essential nutrients. The resistance of the human body is reduced interconnected, as well as the resistance to infectious diseases. Often, the idea of the composition of modern products, their biological value, and the transformations of nutrients in the body cannot be predicted due to insufficient information on their complex interactions.

The factors shown in Figure 3 require a departure from the traditional methods of food production according to the existing similarity and within the framework of existing technologies.

At the end of this chapter, it is necessary to give some quotes from the book of A.Yu. Baranovsky "Dietetics": "... Man is a wonderful creation of nature. In this regard, along with spiritual and physical cultures, a nutritional culture should also take place. I would like to note that the food culture in the deep sense of the word serves as a metabolic control mechanism and should be included in the wider field of physiological, biochemical and other cultures that together make up the biological culture. The latter should take an honorable place along with a moral, social, and other culture.

At present, it is still difficult to characterize all the features of a food culture. But it is now clear that it is primarily associated with an understanding of the activity of the gastrointestinal tract and metabolism in the body. At the same time, national traditions should be taken into account, as certain communities have been formed for a long time and adapted to their diets...

However, it is impossible to state with confidence that traditional food is optimal in all cases...

It is necessary to pay attention to the nutritional culture of young children. Firstly, it should be borne in mind the culture of milk nutrition, and secondly, the ability to modify traditional foods adopted in various social groups, certain communities, so that it matches the physiological needs of a growing organism.

In the light of ideas about food culture, a number of problems should be considered, including the regulation of appetite. Probably, the person has partially impaired that amazing ability to regulate food intake, which is characteristic of animals. Regulation of appetite is one of the important mechanisms for maintaining the constancy of the molecular composition of the body. However, this mechanism is one of the most vulnerable due to a number of circumstances...

Improper nutritional education and improper nutritional behavior of a person, that is, the lack of a food culture, leads to impaired appetite regulation. In this case, the most common error is overeating of certain types of food and malnutrition of others...

The ratio of various components in food and the nature of nutrition, which should ensure the effective functioning of the depot and the "exercise" of various metabolic systems, must be considered from the point of view of food culture...

Now another way is becoming more preferable - the development of adequate food. In this case, polymerization of food (poly- or oligomeric food), the presence of fibrous structures with various properties in it are provided.

In other words, food and nutrition should be adequate to the structurally functional characteristics of the organism and their features. The concept of adequacy allows one to optimize nutrition in accordance with a person's age, the nature of his activity, climatic conditions, etc.

Each era raises the eternal question of how to meet one of the most pressing human needs - the need for food. But the answer to this question is different and depends on the level of our knowledge. A new, deeper and more evolutionary theory of adequate nutrition, formulated on the basis of scientific progress and the development of modern natural sciences, opens up great opportunities and deprives us of a number of illusions, which include ideas about ideal food and ideal nutrition...".

CONCLUSION OF THE CHAPTER

Globalization, industrialization and urbanization as a factor of modern reality have influenced the authenticity of human nutrition. This has led to various risks due to insufficient production of essential substances. The universalization of food preparation processes can lead to the development of unsafe substances that can migrate along the food chain. Regular consumption of fast food products leads to a uniform supply of nutrients that are significantly different from natural ones. Refined foods are depleted in essential substances that can support the human immune system. That is why, even in economically prosperous countries, there is an acute issue of latent hunger, which provokes of nutritionally dependent diseases.

Obviously, global nutrition technologies require the development of special approaches that take into account the accumulation of essential components and contribute to the preservation of the health of generations with regular consumption. Such issues can be resolved by uniting specialists developing food products of a new generation, taking into account the transformation of nutrients at the production and storage process, as well as the metabolism.

REFERENCES

Andrew, C. (2009). *Chop Suey: A Cultural History of Chinese Food in the United States*. Oxford University Press.

Baranovsky, A. Y. (2012). Dietetics (4th ed.). Academic Press.

FAO. (2015). *The State of Food Insecurity in the World*. FAO.

FAO. (2019). *FAO framework for the Urban Food Agenda*. Rome. doi:10.4060/ca3151en

FAO. (2019). *FAO's work on the right to food*. https://www.fao.org/right-to-food/resources/resources-detail/ru/c/1238127/

Guijt, J., Wigboldus, S., Brouwer, H., Roosendaal, L., Kelly, S., & Garcia-Campos, P. (2021). *National processes shaping efforts to transform food systems – Lessons from Costa Rica, Ireland and Rwanda*. FAO.

Guryeva, M.A. (2015). Global problems of world economic development: energy, raw materials, food, demographic and environmental. *Theory and Practice of Social Development, 15*, 34–41.

Ilyin, I. V., & Leonova, O. G. (2016). Research of global processes: Achievements, problems of development and prospects. *Century of globalization*, (1-2), 17–18.

Khisamutdinov, I. A. (2008). Fundamentals of economics and market theory: textbook. Ufa.

Kovalev, E. (2005). New aspects of the world food problem. *World economy and international relations.*, *3*, 3–9.

Kuskova, E. V. (2011). *The problem of hunger and malnutrition in the modern world.* http://www.inspp.ru

Lysak, M.A. (2013) Food problem and ways of its solution in the world. *Fundamental Research, 10*(7), 1545-1549.

Maksakovsky, V. P. (2011). *Global food problem and its geographical aspects*. http://geo.1september.ru

Mcdougall, J. (2021). Globalization of Sichuan hot pot in the "new era". *Asian Anthropology*, *20*(1), 77–92. doi:10.1080/1683478X.2020.1779970

Monteiro, C. A., Levy, R. B., Claro, R. M., de Castro, I. R., & Cannon, G. (2011). Consumption of ultra-processed foods and likely impact on human health. Evidence from Brasilia. *Public Health Nutrition*, *14*(1), 5–13. doi:10.1017/S1368980010003241 PMID:21211100

Moubarac, J. C., Martins, A. P., Claro, R. M., Levy, R. B., Cannon, G., & Monteiro, C. A. (2013). Consumption of ultra-processed foods and likely impact on human health. Evidence from Canada. *Public Health Nutrition*, *16*(12), 2240–2251. doi:10.1017/S1368980012005009 PMID:23171687

Mozaffarian, D., Hao, T., Rimm, E. B., Willett, W. C., & Hu, F. B. (2011). Changes in diet and lifestyle and long-term weight gain in women and men. *The New England Journal of Medicine*, *364*(25), 2392–2404. doi:10.1056/NEJMoa1014296 PMID:21696306

Popkin, B.M., & Adair, L.S. (2012). *Global transition in nutrition and the pandemic of obesity in developing countries.* Academic Press.

Revenko, L. S. (2015). World commodity markets: trends of the XXI century. *Bulletin of St. Petersburg University. Economics,* (3), 27–45.

Sigidov, Y.I., Shol, V.V., Martynenko, E.V., & Barannikov, A.A. (2014). International historical and political analysis of the issue of ensuring food security in the context of globalization. *Scientific Journal KubSAU, 98*(4), 1297–1307.

Ugolev, A. M. (1987). *Natural technologies of biological systems. Academic Press.*

Ugolev, A. M. (1991). *Theory of adequate nutrition and trophology. Academic Press.*

UN: The problem of hunger in the world is exacerbated. (2009). *Air Force.*

Veselov, Y.V., & Junkai, J. (2016). Food globalization processes: the mutual influence of the cultures of the West and the East. *Health and Education in the XXI Century, 18*(9), 135–141.

Vigel, N. L. (2016). Food culture as a method of studying sociocultural transformations. Scientific Journal Service Plus, 10(3), 62–67.

Vlasov, V. A. (2012). Food security issues in the context of globalization (brief historical-political and international legal analysis). *Bulletin of KrasSAU, 11*, 206–212.

Ziyavitdinova, N.M. (2016). Global food problem and ways to solve it. *Young Scientist, 13*, 424–426.

Zobel, E. H., Hansen, T. W., Rossing, P., & Bernt, J. v. S. (2016). Global changes in food supply and the epidemic of obesity. *Current Reports on Obesity, 5*, 449–455.

Chapter 2
Nutrition of the Modern Human and Development of Pathological Conditions

ABSTRACT

This chapter is devoted to the influence of modern man's nutrition on the development and growth of nutritionally dependent diseases. The authors proposed a definition of global food products. The views and opinions of world scientists on the classification of food products from the point of view of safety and possible impact on the human body are presented. The current global food system is a global health crisis. The growing volume of fast-food production and consumption provokes the development of nutritionally dependent diseases. On the example of statistical data on the health status of the adult population, the dynamics of nutritionally dependent diseases over 11 years in Russia have been analyzed. The factors leading to the development of some of nutritionally dependent diseases are noted. The purpose of the chapter is to determine the relationship between the increase in consumption of fast food and the development of civilization disease.

INTRODUCTION OF THE CHAPTER

Epidemiological studies of the last three decades indicate a sharp slowdown in progress in increasing the life expectancy of the population of different countries (primarily industrialized ones), as well as a significant deterioration

DOI: 10.4018/978-1-7998-8197-1.ch002

in health status and a significant increase in the incidence of such diseases, which were much less common 40 years ago.

Experts of the World Health Organization (WHO) believe that health status is determined by the following factors:

- an individual way of life - by 50% (including nutrition about 80%);
- heredity - by 20%;
- environmental conditions - by 20%;
- the work of doctors – by 10%.

Nutrition is the dominant factor of health status. A conscious attitude towards health maintenance in the 21st century begins to take shape in most people, but the united concept of a food culture in the globalization age has not yet been formed for a number of reasons.

Scientific and technological progress has its positive and negative sides. Human life is changing very quickly, becoming more and more intense. The natural mechanisms inherent in the human body do not have time to adapt to new conditions and begin to malfunction, that is expressed in a wide variety of diseases.

Since the 1980s, the national food systems of many developed and post-Soviet countries have been formed by dominant international economic policies aimed at encouraging capital inflows and the rapid trade expansion. The result of this process were international and global trade agreements that allowed transnational food producers, retailers, fast food chains and related corporations to become major players in the food market (Rayner G.et al., 2006, Hawkes C., 2010, Monteiro C.A.et al., 2012,2013).

As mentioned in Chapter 1, transnational corporations form the global food system; they are world producers whose enterprises are based on super-profitable, aggressively advertised products, which today can be called global food products.

Like our colleagues from other countries, we consider global food products as ultra-processed (ultra-transformed) food products and drinks, the production technologies of which are identical around the world.

Global food products are characterized by common properties:

- these are industrial products of a high degree of processing, quick (instant) cooling and consumption, with a long shelf life, recognizable in any country in the world and produced by the same technological methods and techniques from identical raw materials, often by deep processing of agricultural products;

- they have a high calorie content due to the digestible refined food substances contained in them: free oligosaccharides, modified starches, dietary fiber and transformed fats;
- these products are poor in high-grade protein, heme iron, natural (native) vitamins and minerals;
- they contain substances for extending the shelf life of products and maintaining them in an acceptable sensory state over the shelf life: structure-forming agents, preservatives, flavoring agents, coloring agents, etc.;
- food substances in the process of technological processing underwent a deep transformation, as a result of which new products of interaction are formed.

Thus, in the opinion of the authors of this book, global food products are **products designed to quickly hunger satisfying, economically and technologically feasible, adapted and standardized for production and consumption in any country in the world.**

The purpose of the chapter is to find the relationship between the increase in consumption of fast food and the burden of disease of civilization.

MAIN FOCUS OF THE CHAPTER

2.1 The Global Food System and Nutritionally Dependent Diseases

The relation between the global food system and the rapid growth of nutritional-dependent diseases around the world has not yet been studied well enough, although there are more and more works devoted to this global problem in scientific periodicals.

In industrialized countries, the growing number of chronic diseases has become an urgent public health problem, which is further exacerbated by population aging. Risk factor associated with nutrition is one the main causes of the spread of chronic diseases in the WHO European Region (Lim SS et al., 2012).

Fast-food habits are changing rapidly in low- and middle-income countries. The available data indicate that there is a process of crowding out traditional, centuries-old food products and freshly prepared dishes with global food products, which increases the risk of nutritionally-dependent diseases and contributes to an increase in the number of these diseases (Hughes R. et al., 2005).

Obviously, nutrition is an important public health policy tool, as it is a determinant of health that can be influenced by primary prevention measures.

In response to the current situation, governments around the world are introducing strategies and policies that include comprehensive interventions to improve nutrition (Liu PJ et al., 2014; Serra-Majem L., 2009). The most popular measure is food products marking on the front of the package (Hercberg S., 2017; Chantal Julia, Serge Hercberg, 2017; Di Ciaula A., Portincasa P. 2019).

British Food Standards Agency nutrient profiling system (FSA-NPS) is validated in the United Kingdom food database. Recently, the French government announced the introduction of a new food labeling system on the front of the package at the national level. This system, the so-called Nutri-Score scale, is a five-color marking developed by the State Research Association "Research Group on Nutrition Epidemiology" (Leonardi A et al., 2017).

There is an alternative food classification system, called NOVA, which focuses on the nutritional composition of the diet, takes into account the degree of processing, the presence or absence of additives, etc. According to NOVA, food can be divided into 4 groups:

1. Unprocessed or minimally processed foods are natural foods, such as herbs, berries, seeds, milk, eggs, etc.
2. Processed natural food products that have passed certain stages of the technological processes: drying, crushing, grinding, fractionation, filtration, heat cooking, fermentation, pasteurization, cooling, distillation, freezing, packaging. These processes are designed to bring natural food products to the state of their culinary readiness, characterized by safety, improved structural, mechanical and organoleptic properties, to create conditions for safe short-term storage.
3. Processed food products. These are products that have passed, basically, the stage of preservation: salted, pickled products, jams, etc.

Processes include various methods of food preserving or cooking, and, in the case of bread and cheese, the method of non-alcoholic fermentation. Most processed foods have two or three components, and are also recognizable as modified versions of the 1st product group. In this case, the purpose of the processing is to increase the shelf life of food of the 1st product group, to diversify or increase the taste.

4. Ultra-processed or ultra-transformed food products – soft drinks, sweet or salty snacks, processed meat products and pre-prepared frozen foods, not modified products, but produced from secondary substances obtained from foods and additives with a low content of the 1st product group. The ingredients of these formulations also include those used in processed foods, such as sugar, butter, fats or salt. But ultra-transformed foods also include other sources of energy and nutrients, usually not used in culinary products and dishes. Basically, these are products of deep processing of agricultural products, such as casein, lactose, whey, gluten and others.

Many of these products are derived from further processing of food components such as hydrogenated or trans esterified oils, hydrolyzed proteins, soy protein isolate, maltodextrin, invert sugar, and high fructose corn syrup.

Supplements in ultra-transformed foods include some of them, and are also used in processed foods, such as preservatives, antioxidants, and stabilizers. Classes of additives containing only in ultra-transformed products include supplements that are used to imitation, intensify eating experience or mask the unattractive organoleptic aspects of the final product.

These additives include dyes, color stabilizers; flavors, flavor enhancers, sugarless sweeteners and other auxiliary substances.

Many sequential processes are used to combine a large number of ingredients and create the final product (hence the "ultra-transformed"). The processes include several process steps, such as hydrogenation and hydrolysis, extrusion and molding, and pretreatment for frying.

The overall goal of ultra-transformation is to produce branded, convenient (durable, ready-to-eat), attractive (hyper-palatable) and highly profitable (low-cost) food products designed to displace all other groups of products. Typically, ultra-transformed foods are attractively packaged.

The constantly growing production level and consumption of these products is a global crisis in health conservation, affecting all segments of the population and the state of the world causing "diseases of civilization": cardiovascular diseases (hypertension, myocardial infarction, stroke); gastrointestinal diseases (gastric ulcer, pancreatitis, hepatitis, dysbacteriosis, dysfunction of the small and large intestines); endocrine diseases (diabetes, thyrotoxicosis); oncological diseases; allergy; metabolic diseases (atherosclerosis, overweight and obesity).

These diseases affect not only an increasing number of people of working age, but also very young people, causing enormous economic damage to states. So, signs of atherosclerotic vascular changes began to be detected even in children of primary school age. Against the background of physical inactivity developing, high-calorie diet causes a significant reduction in life expectancy.

Currently, the problem of extremely low provision of the population with the majority of vitamins, microelements, dietary fiber is relevant; deficiency of iodine, selenium, iron, calcium is widespread.

A person of a modern urbanized society with traditional nutrition is essentially doomed to certain types of food deficiency, which is accompanied by the inability of the corresponding protective systems of the body to adequately respond to adverse environmental influences, which in turn sharply increases the risk of many diseases.

Any health problem is biochemical changes in the cells.

Today, science studies the safety of anthropogenic impact from the outside, microbiological changes and legislative regulation of permissible levels of contaminants. New technologies, the consequences of the widespread popularization of global food products are ahead of research on the emergence of new risks of negative effects on the body at regular consumption of these products. Thus, the modern approach to health maintenance is based on a subsequent statement of facts that have already happened and the number of nutritionally-dependent noncommunicable diseases.

For example, widespread toxicants, the effects of which are being actively studied at present (mainly bisphenol A, phthalates, pesticides) that migrate to modern food products from the soil during cultivation and from packaging during storage, destroy the endocrine system (Rochester JR., 2013). Scientists continue studies that reveal the association of chemicals with serious health problems.

Recent studies show that bisphenol A causes the development of chronic conditions such as diabetes and heart disease, and also harms the fertility of men and women. It inhibits sperm motility by decreasing ATP levels in sperm cells, as well as damaging proteins that facilitate fertilization and embryonic development. BPA also negatively affects the female genital organsand adversely affects the reproductive cycle. In addition, recently it became known that bisphenol A has a harmful effect on the body even in very low doses (Di Ciaula A., 2019; Baralić K. et al., 2020).

Phthalates also have adverse health effects. Their exposure is associated with DNA damage in human semen and a decrease in testosterone levels, which can lead to infertility. Recent studies also show that phthalates impair cognitive development and contribute to allergic disorders in children.

Both bisphenol A and phthalates have epigenetic effects on health. They cause changes in gene expression that can be transmitted across several generations. Thus, exposure to bisphenol A in parents can cause health problems in their children and possibly even their grandchildren. The above toxicants can stimulate obesity in children and adults, mainly acting on the differentiation pathway, connecting multipotent stromal stem cells with mature adipocytes, modulating epigenetic factors and influencing a number of mechanisms that ultimately lead to changes in eating habits, increased adipocyte formation and fat storage (Blair A. et al., 2014; Souza GDS et al., 2017; Nava-Castro K.E. et al., 2019).

Environmentally stable pesticides enter the human body with food in 95% of cases, with water in 47%, and only in 0.3% of cases with atmospheric air through the respiratory tract and through the skin. Pesticides, entering the human body in large doses, affect almost all organs, causing dystrophic changes in tissues, disrupting metabolism, and inhibiting the functions of the central and peripheral nervous system.

Today, the entire population of the Earth is exposed to the chronic effects of small doses of pesticides, and the degree of their impact depends on the diet, geographical location and level of industrial development of the region. For example, the level of pesticides' exposure on people living in areas with actively developed agriculture is much higher than on those who live away from agricultural production facilities.

It was previously believed that "the dose determines the poison". This meant that small amounts of poison do not lead to any consequences, and large doses are fatal. However, studies of scientists have shown that repeated small doses are more dangerous than a single significant dose. This statement is also true for affect of pesticides. Every day, every person receives micro doses of pesticides, which affect the body along with other adverse anthropogenic factors.

Possible activity of the chronic effects of low intensity pesticides is classified as long-term effects. Pesticides can pass down through generations.

Most pesticides are fat-soluble, which means that they can be excreted in breast milk during breastfeeding and, accordingly, can negatively effect on the child's body development. Pesticides can cause mutations (changes in the genetic apparatus of the cell). Pesticides with a pronounced mutagenic effect (diazinon) can cause mutations in germ cells.

In most cases, pesticides are non-genotoxic carcinogens, i.e. they do not directly react with DNA, but are triggers of damage to the genetic material of cells, which leads to the tumor development. For example, organochlorine pesticides can cause breast cancer, sarcoma, leukemia, lymphoma, triazine herbicides can cause ovarian cancer.

Numerous experiments on laboratory animals have shown that organochlorine insecticides (DDT) can cause liver tumors; carbamates (maneb, ziram) induce thyroid tumors, chlorothalonil causes adenomas and kidney cancer; dichlorvos causes tumors of the esophagus and stomach (Filip L. et al. 2004; Forastiere F et al.,2015; Tsatsakis A.M. et al.; 2017, Ji C. et al., 2020).

To date, genetically modified foods (GMOs) have increased yields and reduced pesticide application. Nevertheless, the use of GMO products causes controversy among politicians, scientists and consumers regarding their possible environmental and medical risks (Shephard G.S., 2008; Thibault F. et al., 2018).

It should be noted that many other contaminants of chemical and biological origin have negative effect on the human body. The effect of mycotoxins (in particular, aflatoxin) on the development of liver cancer, suppression of immunity and growth retardation has been studied by many scientists.

It should be noted that despite a large number of studies, they are quite disparate in nature and require a generalized approach both from the side of the need for food components depending on the influencing factors, and knowledge about modern food products from the technological transformations of substances, positively and / or negatively affecting the body.

The ideal option in the daily diet of a person is the consumption of natural foods from an authentic cuisine, replenishing nutrients and essential substances. But even with a full and balanced diet, part of the nutrients may not enter the body for a number of reasons:

- due to stomach diseases;
- lack of certain enzymes;
- “slagging” of the body;
- cooking, etc.

In the modern world, there are a number of nutritional concepts that are based on scientifically based hypotheses. It should be borne in mind that at present there are more than a hundred thousand kinds of food products, and every year new combined and non-traditional food products appear. The amount of food that a person eats, is usually called an "actual nutrition". However, only a certain part of the nutrients entering the body with food can be absorbed by the body during digestion. For many nutrients, the difference between intake and absorption can reach from several to tens of percent. This difference increases in people with digestive diseases. In such cases, a reduced absorption state may occur for certain groups of nutrients, that is called malabsorption.

The following facts should be taken as a basis:

- the nutritional status includes the calculation of the amount of consumed basic food components: proteins, carbohydrates, lipids, mineral components, trace elements, vitamins, water, electrolytes, dietary fiber, as well as a certain list of minor biologically active components that can enter the body in unsplit form and have nutritional or physiological significance;
- the nutritional status is determined by calculation methods, based on the chemical composition of food products, by directly monitoring the nutrition of a given person for a certain period of time or by polling. Knowing the chemical composition of food products, it is relatively easy to calculate the amount of certain food substances that enter the body. It should be noted that a tabular calculation of the chemical composition of food products may differ significantly from the actual content of these components in the product. This applies to almost all food products, but especially to products of plant origin, the macro- and microelement composition of which depends on the geochemical composition of the soil, climate, level of solar insolation, storage and processing conditions:
- when evaluating and analyzing various nutritional concepts, it is also necessary to take into account the proportion of absorption of nutrients and factors affecting this process.
- Most nutrients are always absorbed in the digestive tract than they enter the blood and lymph, because in the process of digestion, nutrients are recycled in the intestines, and only a part of them is included in the metabolism. Another part will be consumed by intestinal microorganisms, parasites (protozoa, helminths), and, without being absorbed, will be excreted with feces. These losses can be taken into account by introducing a correction in the form of an assimilation coefficient. At the same time, the digestibility of nutrients

> depends on many factors: the composition of the diet (the ratio of various components and their amount), the state of digestive function, intestinal microbiocenosis, the content of food substrates in the body, and a number of other reasons, therefore, the nutrient assimilation rate is a variable.

In the gastrointestinal tract, a significant part of food substrates produced or transformed by intestinal microorganisms is absorbed by the body. These include vitamins, volatile fatty acids, as well as lipids, amino acids, carbohydrates, nucleotides and other substances that are formed as a result of the vital activity and degradation of bacteria. Some nutrients are absorbed by the body during digestion of dietary fiber by microorganisms. All this nutrients formed in the gastrointestinal tract can be called "endogenous microbial nutrition".

The difference between the nutrients entered in the gastrointestinal tract and those absorbed during digestion is the loss of nutrients.

The balance of nutrients is a variable quantity, although the body seeks to regulate and maintain it at a constant level. With an excess of certain food substances, their absorption and utilization reduce, their excretion enhances. On the contrary, with an insufficient content of these nutrients in the diet, the opposite effect occurs. With a significant excess of food substrates in the diet, the balance can be disturbed causing accumulation diseases. One of the most common of them is obesity. In the case of a significant reduction in the content of food substrates in the diet, the balance of nutrients can also decrease and establish at a lower level, which leads to diseases of nutritional deficiency.

It is important to emphasize that rational nutrition for each person is not a certain constant value. On the contrary, rational nutrition is a variable, it changes with age, and depends on the gender, ethnicity of a person, the level of physical and psycho-emotional activity, health status, as well as on external factors. A balanced diet is a certain amount and ratio of nutrients in the composition of nutrition, which contributes to the normal and stable functioning of metabolic processes and the body.

Nutrition should be balanced. This means that in order to achieve a beneficial effect, all the main nutrients must be in a certain ratio (be balanced). Not only a deficiency, but also an excess of the main groups of food substances (proteins, fats and carbohydrates), and even of individual nutrients (amino acids, trace elements, vitamins) can lead to a disorganization of metabolic processes in the body.

Various proportions of the content of nutrients in the diet were calculated and experimentally verified, the optimal ratio were determined. However, these ratios are generalized and are also a guideline for calculating the individual nutrition of each person. They must be constantly adjusted when changing lifestyle, health status and external conditions.

The Russian physiologist A.M. Ugolev introduced the concept of "adequate nutrition", which complemented the concept of rational and balanced nutrition. He emphasized the important role of the ballast components of food (dietary fiber) and intestinal microflora as an additional source of food substrates for each person. According to the author, adequate nutrition is nutrition that takes into account the importance of dietary fiber and intestinal microflora in providing the body with an additional amount of food substrates (minus those that are utilized by the bacteria). In our opinion, the concept of adequate nutrition in its meaning is closer to individual or optimal nutrition.

Optimal nutrition is a nutrition that takes into account the needs of a particular person in a given period of time.

Nutrition optimization is the ultimate goal in calculating any nutrition. However, it is quite difficult to take into account individual human nutritional needs. At the same time, it must be borne in mind that by improving nutrition in one way, for example, protein or vitamin C, we optimize the functioning of certain metabolic systems of the body. At the same time, for other body systems, this level of nutrients will not be optimal. Therefore, the concepts of "optimality" or "adequacy" are relative concepts. It would be more correct to talk about the appropriateness of nutrition, which should help achieve a certain goal in regulating the metabolism in the body and ensure the stable operation of all systems.

Adequate and conscious nutrition is still ideal for most people, as industrial nutrition and a sharp increase in global food consumption are becoming dominant in many regions of the world. It should be noted that global food products are microbiologically safe, convenient, very tasty and affordable due to unified technological processes. Surveys in Europe, the USA, Canada, New Zealand, and Brazil that measure individual food consumption, household spending on food, or sales in supermarkets have shown that global food makes up from 25 to 50% of total daily energy consumption (Muncke J., 2010).

This nutritional trend is alarming and worth exploring. According to many researchers (Muncke J., 2010), in global products that underwent deep heat treatment, there are substances with carcinogenic properties (acrylamide, heterocyclic amines and polycyclic aromatic hydrocarbons) (FerlayJ., Soerjomataram I., Dikshit R., et al., 2012). A factor of nutritional hazard is also storage with subsequent heat treatment of packaged ultra-processed foods. This is due to the intensification of the migration of contaminants from the package, especially into the fat phase of the product, materials contacting with food products (bisphenol A, etc.).

2.2 Dynamics of Nutritionally Dependent Diseases. Factors, Development Trends

Exploring the potential health effects of ultra-processed foods is a very recent area of research.

The human body can not assimilate many modern products, this negatively affects the life quality of a modern person.

The authors analyzed statistical data on the registration of diseases of the adult population, provoked by the nutritional factor. The analysis was carried out according to generalized studies in the Russian Federation with the most complete picture in various categories[1], as well as according to some world statistics available for analysis.

The incidence of the population of the Russian Federation as a whole and for diseases with an alimentary factor, are presented in Table 1. The data are examined in the context of seven years (from 2010 to 2016).

Various diseases can be associated with nutritional factors, especially at the violation of nutritional status, lack of essential substances, as well as regular consumption of fast food products.

It should be noted a significant increase in nutritional-dependent diseases. Neoplasms in the Russian Federation for the period from 2010 to 2015 increased by 13.2%, while over the next year by 18.5%, that is, an increase in 2010 - 2016 amounted to 18.5%.

The winner of the International Oncology Prize A. Zhuayo writes in "Nutrition and Cancer" that malnutrition leads to obesity and cancer, "... 50% of all cancers present today and those that will develop after 2000 are the result of improper nutrition ...".

Cancer is a great worldwide problem; according to the World Cancer Research Foundation / American Cancer Research Institute, 14.1 million new cases were diagnosed in 2012 (Latino-Martel P., Cotte V., Druesne-Pecollo N., et al., 2016), about a third of the most common neoplasms could be avoided by changes in lifestyle and eating habits in developed countries (Latino-Martel P. et al., 2016).Therefore, achieving a balanced diet should be considered as one of the most important modifiable risk factors in primary cancer prevention(Canella DS. et al., 2014).

Table 1. The incidence of the population of the Russian Federation as a whole and for diseases with an limentary factor

Disease	2010	2015		2016	
	In total	In total	Growthto2010, %	In total	Growthto2010, %
In total	226159.6	234331.7	3.6	237067.7	4.8
Neoplasms	5847.0	6616.4	13.2	6927.6	18.5
Diseases of the blood,	1741.5	1825.5	4.8	1975.5	13.4
blood-forming organs and certain disorders involving the immune mechanism: In total among them anemia	1570.5	1631.8	3.9	1637.8	4.3
Endocrine system diseases, nutritional disorders, metabolic disorders In total	8494.4	10872.2	27.9	11457.8	34.9
among them diabetesmellitus	3378.0	4418.3	30.8	4610.8	36.5
obesity	1161.7	1726.9	48.6	1825.4	57.1
Diseases characterized by high blood pressure	11787.2	13948.9	18.3	14545.5	23.4
Digestive system diseases In total	16069.5	17132.8	6.6	17206.6	7.0
among them gastric and duodenal ulcer	1495.7	1275.9	- 14.7	1244.3	- 16.8
Congenital malformations, deformations and chromosomal abnormalities	991.3	1096.7	10.6	1151.3	16.1

** Healthcare in Russia. 2017: statistical book / Rosstat.Moscow, 2017*

Nevertheless, the epidemiological data linking the consumption of ultra-processed foods with the risk of disease is still very scarce and is mainly based on cross-sectional and environmental studies (Julf&Hemmingsson, E., 2015; Luzada M.L, et al.,2016; Basiak-Rasała A., et al., 2019).

T. Fiolet et al. conducted a prospective study to assess the relation between the consumption of ultra-processed foods and the cancer incidence. The authors' conclusions were based on a large cohort study with a detailed and relevant dietary assessment. The results of the study showed that an increase in the proportion of ultra-processed foods in the diet by 10% led to a significant increase in the risk of developing general and breast cancer (by more than 10%).

Over the last year of the examination, there has been a significant increase in blood and blood-forming organs diseases and certain disorders involving the immune mechanism, including anemia, although it was not as significant as the development of the neoplasm.

The largest changes in the dynamics of diseases of the endocrine system, eating disorders, metabolic disorders amounted to 27.9% for the period from 2010 to 2015, and to 34.9% for the period from 2010 to 2016. Moreover, a catastrophic increase in this group of diseases is observed for diabetes mellitus: 30.8% from 2010 to 2015 and 36.5%for the period from 2010 to 2016.

The World Health Organization predicts that diabetes will be the seventh leading cause of death in 2030. Most people with diabetes suffer from type 2 diabetes (T2DM), which can mostly be avoided.

The most important modifiable risk factors for developing type 2 diabetes are overweight and obesity, unhealthy diets, a sedentary lifestyle, and smoking. Even in a pre-diabetic state, improved diet can slow or even stop the progression of this disease.Studies by A. Basiak-Rasała et al. (Basiak-Rasała A. et al., 2018) indicate an inverse relation between the consumption of fresh vegetables and fruits, whole grains, low-fat dairy products, fish, nuts, and the risk of type 2 diabetes. Food groups that increase the risk of type 2 diabetes include: red and processed meats, refined grains, and sugar-sweetened drinks.

U. Masharanietet al. (Masharani U. et al., 2015) investigated the conditions of patients with type 2 diabetes mellitus and indicate the beneficial effect of the paleontological diet, even with short-term use. Patients have improved glucose and lipid profiles compared to a regular diet containing moderate salt intake, low-fat dairy products, whole grains and legumes.

New data on the influence of food on the spread of diabetes were obtained by Chinese scientists (Castro I. et al., 2015), studying the associations of red meat, poultry and fish with cardiometabolic diseases. According to a study, consumption of red meat and fish, but not poultry by adult Chinese, showed a positive association with the risk of diabetes, especially among urban participants.

Over the analyzed period, the number of diseases characterized by high blood pressure increased by 23.4% in the Russian Federation. A substantial growth (5.1%) was in the last analyzed years (2015 and 2016). According to the World Health Organization, arterial hypertension (AH) is the leading cause of death and a reason of approximately 40% of deaths from cardiovascular disease, chronic kidney disease (CKD), and type 2 diabetes mellitus (T2DM) (Heinicke V. et al., 2020).

It is known that about 80% of all deaths from cardiovascular diseases could have been prevented if a healthy diet were observed. A commitment to a healthy lifestyle throughout life can significantly reduce the risk of cardiovascular disease by up to 70%, therefore it is a key component in the primary prevention of cardiovascular disease. Cardioprotective nutrition (a high proportion of unsaturated fatty acids, low amounts of saturated fatty acids and low salt intake) plays a special role in this case (Baldacci S. et al., 2018).

It should also be noted the growth of congenital anomalies (malformations), deformations and chromosomal abnormalities up to 16.1% in the Russian Federation. Every year, about 7.9 million children in the world are born with congenital malformations. Congenital anomalies (CA) are one of the main causes of fetal death, infant mortality and incidence, as well as of long-term disability. Congenital anomalies have been the subject of systematic registration activities for a long time in many geographical areas of Europe and around the world. They are often associated with disabilities of various types and severity, including in developed countries. According to the World Health Organization (WHO), approximately 300.000 newborns diagnosed with congenital defects die within the first 28 days of life. In Europe, congenital anomaly is the leading cause of perinatal mortality: European network of population-based registries for the epidemiological surveillance of congenital anomalies (EUROCAT) estimated perinatal mortality by CA at 9.2 per 10.000 births in 2008-2012. In addition, approximately 25% of infant mortality is attributable to CA and about 50% of infant mortality is attributable to perinatal incidence of prenatal origin. Regarding long-term survival, a recent population-based study conducted between 1985 and 2003 in the UK showed that 20.5-year survival is 85.5% of children born with at least one CA.

According to the Centers for Disease Control and Prevention, approximately 3.3% of live births in the United States have a severe congenital defect. Since CA is a significant public health problem, an effective primary prevention strategy should be a priority for government policy and the health system. It has been suggested that CA may develop during the first trimester of pregnancy as a result of hereditary polygenic defects or the interaction of a gene with the environment. The etiology is predominantly multifactorial; it is caused by complex interactions between genes and the environment, which modify the normal embryonic-fetal development, especially at the stage of organogenesis. In particular, environmental factors (for example, chemical toxicants, infectious agents, maternal diseases and exogenous factors) can have a preconception mutagenic and a postconception teratogenic effect, epigenetic effect and cause endocrine disorders. Regarding the genetics of the cause, there are genetic chromosomal aberrations or dysgenesis. In addition, socioeconomic factors affect reproductive health by differentiating the effects of other risk factors.

In recent years, the importance of the environment as one of the main factors of reproductive risk has been emphasized. A person may be exposed to pollutants in the workplace, and the population may be exposed to numerous sources of environmental pollution - water, soil and air. Pregnant women and the developing fetus are especially sensitive to environmental influences (Baldacci S. et al., 2018).

A study conducted by S. Bellizzi et al. (2016) shows that maternal chronic hypertension poses a significant risk of congenitalanomalies of the kidneys, limbs and lips / crevices / palate in newborns, and the risk is further exacerbated by eclampsia (Bellizzi S. et al., 2016).

Based on the analysis of general data on the incidence of the population of the Russian Federation as a whole and on diseases with nutritional factors, it should be noted that obesity and diabetes are the most progressive diseases in recent years.

Table 2 presents statistics on registered malignant neoplasms in patients with a diagnosis established for the first time in their lives. The analysis is presented by age and gender. The 24.87% growth of malignant neoplasms registered for the first time in 11 years for all age groups is obvious. The highest growth is observed in 0-14 years old men and women: by 37.20% for men and 39.04% for women. These data, indicating an unfavorable situation for malignant diseases in children, suggest that the nutrition of mothers and children does not meet the necessary level to ensure health. However, it is impossible to assess the specific effect of the totality of nutrients on ontogenesis.

Table 2. The incidence of malignant neoplasms by gender and age groups (registered diseases in patients with a diagnosis established for the first time in their life)

Group	2005	2010		2013		2014		2015		2016	
		Number	%	Number	%	Number		Number	%	Number	%
In total, persons											
men	219093	237982	8.62	245180	11.91	259425	18.41	270046	23.26	273585	24.87
women	250102	278892	11.51	290707	16.24	307545	22.97	319335	27.68	325763	30.25
including at the age, year:											
0-14 men	1301	1565	20.29	1594	22.52	1643	26.29	1684	29.44	1785	37.20
women	1081	1198	10.82	1293	19.61	1426	31.91	1500	38.76	1503	39.04
15-19											
men	920	596	-35.22	506	-45.00	558	-39.35	565	-38.59	566	-38.48
women	805	605	-24.84	498	-38.14	512	-36.40	551	-31.55	529	-34.29
20-39											
men	6826	7704	12.86	7814	14.47	8087	18.47	8568	25.52	8527	24.92
women	13052	15365	17.72	16529	26.64	17037	30.53	18105	38.71	18613	42.61
40-59											
men	69760	74037	6.13	72014	3.23	72976	4.61	73920	5.96	73118	4.81
women	82892	90418	9.088	90235	8.86	93174	12.40	94557	14.07	95532	15.25
60 and											
older men	140286	154080	9.83	163252	16.37	176161	25.57	185309	32.09	189589	35.14
women	152272	171306	12.50	182152	19.62	195396	28.32	204622	34.38	209586	37.64

* *Healthcare in Russia. 2017: statistical book / Rosstat.Moscow, 2017*

Women of childbearing age (20-39 years) are in groups with intensive growth of newly diagnosed oncological diseases. This group showed an increase of 42.61% compared to 2005. It is assumed the relation between the growth of oncological diseases of women and children, which is correlated with the globalization of the food culture and the loss of the products of oncoprotective action in nutrition.

Another age with intensive growth of newly diagnosed oncological diseases is the age of 60 (more than 35.14% for men and 37.64% for women).

The data presented in Table 3 also confirm the increase in the incidence of malignant neoplasms.

Table 3. The dynamics of the incidence of malignant neoplasms

	2005	2010		2013		2014		2015		2016	
		Number	%	Number	%	Number	%	Number	%	Number	%
Registered diseases in patients with a diagnosis established for the first time in their life											
In total, thousand persons	469.2	516.9	10.2	535.9	14.2	567.0	20.8	589.4	25.2	599.3	27.73
The number of patients registered in medical organizations (at the end of the year)											
In total, thousand persons	2386.8	2794.2	17.0	3098.9	29.8	3291.0	37.8	3404.2	42.63	3518.8	47.43
From the number of registered patients with malignant neoplasm, established for the first time in life											
The fourth stage neoplasm In total, thousand persons	103.3	106.7	3.3	108.8	5.32	112.7	9.10	114.7	11.04	117.3	13.5

* *Healthcare in Russia. 2017: statistical book / Rosstat.Moscow, 2017*

It should be noted the increase in the number of people with the fourth stage neoplasm diagnosed for the first time from 3.3% (in 2010 compared to 2005) to 13.5% (in 2016 compared to 2005).

It is of interest to consider a comparative analysis data on the number of oncological diseases.

The increase in the number of registered diseases in patients with a diagnosis established for the first time in their lives in the urban population is slightly lower than in the rural one (28.45 and 25.4% in 2016 compared to 2005). This trend is variable over the years. So, in 2013, the number of registered oncological diseases in the rural population was slightly higher than in the urban population (Table 4).

If we analyze the data per 100,000 people, then another trend is observed - the percentage of incidence in the rural population is either equal to or higher than in the urban one.

Analyzing data on the population registered in medical institutions, we can note almost the same percentage between urban and rural population.

Tables 2, 3, and 4 show a steady increase in the number of malignant tumors, the absence of a significant difference between the number of diseases in urban and rural areas, as well as risk groups with the highest incidence rate.

Table 4. Comparative analysis of the incidence of malignant neoplasms in urban and rural population

	2005	2010		2013		2014		2015		2016	
		Number	%	Number	%	Number	%	Number	%	Number	%
Registered diseases in patients with a diagnosis established for the first time in their life: In total, thousandpersons											
Urban population	357.8	396.0	10.68	407.6	13.92	432.4	20.85	451.0	26.05	459.6	28.45
Including: men	162.5	177.2	9.05	180.6	11.14	192.7	18.58	200.5	23.38	203.5	25.23
women	195.3	218.8	12.03	227	16.23	239.6	22.68	250.5	28.26	256.1	31.13
Rural population	111.4	120.9	8.53	128.3	15.17	134.6	20.83	138.4	24.24	139.7	25.40
Including: men	56.5	60.8	7.61	64.6	**14.34**	66.7	18.05	69.6	23.19	70.1	24.07
women	54.8	60.1	9.67	63.7	16.24	67.9	23.91	68.8	25.55	69.6	27.01
per 100 000 persons of population											
Urban people	**342.1**	**376.3**	**10.00**	**383.3**	**12.04**	**400.1**	**16.95**	**415.8**	**21.54**	**422.3**	**23.44**
among them:											
men	337.5	368.5	9.19	371.5	10.07	389.9	15.53	404.2	19.76	408.9	21.16
women	343.6	382.8	11.41	393.3	14.46	408.7	18.95	425.6	23.86	433.6	26.19
Rural people	**294.8**	**321.4**	**9.02**	**345.1**	**17.06**	**354.0**	**20.08**	**364.8**	**23.74**	**369.3**	**25.27**
among them:											
men	308.5	338.6	9.76	362.1	17.37	365.5	18.48	381.6	23.70	385.0	24.80
women	271.6	305.7	12.56	329.3	21.24	343.3	26.40	349.2	28.57	354.8	30.63
The number of patients registered in medical organizations (at the end of the year): In total, thousandpersons											
Urban population	1890.0	2225.5	17.75	2479.7	31.20	2611.9	38.20	2691.3	42.40	2782.9	47.24
Rural population	496.8	568.7	14.47	619.1	24.62	679.2	36.71	712.9	43.50	735.9	48.13
per 100 000 persons of population											
Urban population	1806.8	2114.7	17.04	2327.3	28.81	2412.1	33.50	2476.9	37.09	2552.4	41.27
Rural population	1315.3	1512.1	14.96	1667.9	26.81	1788.0	35.94	1881.7	43.06	1948.4	48.13

* *Healthcare in Russia. 2017: statistical book / Rosstat.Moscow, 2017*

The health of women of childbearing age is an important factor in the generation health. Statistical data on the incidence of women by certain types of diseases that may be dependent on the nutritional factor are presented in

Table 5. The increase in cancer in women from 2005 to 2010 was 11.52%, and from 2005 to 2016 – 30. 27%. Malignant neoplasms of the breast prevail among oncological diseases in women: for the analyzed 11 years the growth amounted to 38.38%, while a sharp spike is observed for the period from 2010 to 2016. During these 6 years, the number of diseases increased by almost 23%, compared with the first five years of the analysis - by 15.56%.

Table 5. Incidence of certain diseases in women

	2005	2010		2013		2014		2015		2016	
		Number	%	Number	%	Number	%	Number	%	Number	%
Registered diseases in patients with a diagnosis established for the first time in their life: Intotal, thousandpersons											
Malignantneoplasms	250.1	278.9	11.52	290.7	16.23	307.5	22.95	319.3	27.67	325.8	30.27
Among them: lacteal gland	49.5	57.2	15.56	60.7	22.63	65.1	31.52	66.6	34.55	68.5	38.38
cervix and body of uterine. placenta	30.2	34.6	14.57	37.8	25.17	39.8	31.79	41.2	36.42	42.4	40.40
ovary	12.3	13.1	6.50	13.3	8.13	13.6	10.57	14.0	13.82	14.0	13.82
Infertility	52.5	70.7	34.67	81.0	54.29	83.1	58.29	94.2	79.43	92.8	76.76
Per 100 000 women: Malignantneoplasms	327.2	363.1	10.97	377.3	15.31	392.2	19.87	406.4	24.21	413.9	26.50

* *Healthcare in Russia. 2017: statistical book / Rosstat.Moscow, 2017*

Table 6. Health status of pregnant women, women in labor and women in childbirth

	2005	2010		2013		2014		2015		2016	
		Number	%	Number	%	Number	%	Number	%	Number	%
The number of women who have completed pregnancy: In total, thousand persons, including:											
prematurelabor	44.2	62.4	41.18	78.8	78.28	76.7	73.53	76.9	73.98	73.3	65.84
abort	53.2	66.9	25.75	66.7	25.38	68.2	28.20	68.9	29.51	66.2	24.44
Women with completed pregnancy, suffered with, percent											
diabetesmellitus	0.2	0.4	**100.00**	0.7	**250.00**	1.3	**550.00**	2.2	**1000.00**	3.1	**1450.00**
The number of diseases complicating childbirth (per 1000 births): including:											
diabetesmellitus	1.6	3.7	**131.25**	7.8	**387.50**	14.1	**781.25**	23.7	**1381.25**	31.9	**1893.75**

* *Healthcare in Russia. 2017: statistical book / Rosstat.Moscow, 2017*

Another disease that may be associated with the consumption of global food is infertility.

One of the probable reasons for the sharp increase in infertility among young women may be an increase in the amount of epoxides in food, in particular in oxidized fats. These cyclic ethers, due to the tension of the three-membered ring, are highly reactive in ring-opening reactions. Epoxides are formed in the human body as a result of biotransformation of xenobiotics [50] or enter with food containing an altered fat component. Due to their high reactivity, epoxides easily alkylate the nucleophilic centers of nucleic acids, resulting in changes in the DNA structure, which leads to an increase in the number of mutations.

Over the analyzed period, infertility in women increased by more than 76%, and a sharp increase was observed after 2010.

It is important to note that the number of complications during childbirth also increases (Table 6), provoked by the growth in number of alimentary-dependent diseases.

More and more studies associate the occurrence of such diseases with carbonyl stress caused by the accumulation of active carbonyl compounds, including the consumption of foods with a high content of melanoidins, lipid oxidation products

It should be noted that during the period from 2005 to 2016, the percentage of pregnant women with diabetes mellitus reached a catastrophic figure of 1450.00%. And the number of women in whom this disease caused a complication during childbirth increased by 1893.75%. Moreover, the growth from 2010 to 2016 amounted to more than 10 times. Such an increase in diseases associated with the nutritional factor can also cause the teratogenic factor in future generations.

CONCLUSION OF THE CHAPTER

The authors of this work state the fact of a significant effect of fast food products on human health. Along with the increase in consumption of fast food, the number of disease grows rapidly. According to many studies, the authors of the book can declare that nutrition is a critical factor in the long term direct impact on human health.

Currently relevant are the directions in the development of specialized food products, as well as personalized nutrition products and diets (based on the allelic polymorphism of individual "predisposition genes" associated with the absorption of certain food nutrients), aimed at improving the nutritional structure of the general population and individual groups.

When developing personalized (individual) diets, the following factors should be considered:

- the need of the individual for nutrients, depending on the condition, the load performed, the stress factor;
- the body's ability to absorb nutrients, that is, the enzyme and microbiotic activity of the body, in combination with the state of the digestive system and the immune system;
- individual preferences with adequate product interchangeability.

REFERENCES

Baldacci, S., Gorini, F., Santoro, M., Pierini, A.,Minichilli, F. & Bianchi, F. (2018). Environmental and individual exposure and the risk of congenital anomalies: a review of recent epidemiological evidence. *Epidemiol Prev., 42*(3-4), 1–34.

Basiak-Rasała. (2019). *Roczniki Panstwowego Zakladu Higieny*. Academic Press. PMID:31960666

Baralić, K., Buha Djordjevic, A., Živančević, K., Antonijević, E., Anđelković, M., Javorac, D., Ćurčić, M., Bulat, Z., Antonijević, B., & Đukić-Ćosić, D. (2020). Toxic Effects of the Mixture of Phthalates and Bisphenol A-Subacute Oral Toxicity Study in Wistar Rats. *International Journal of Environmental Research and Public Health, 17*(3), E746. Advance online publication. doi:10.3390/ijerph17030746 PMID:31979393

Bellizzi, S. (2016). Are hypertensive disorders in pregnancy associated with congenital malformations in offspring? Evidence from the WHO Multicountry cross sectional survey on maternal and newborn health. *BMC Pregnancy Childbirth, 16*(1), 198. doi:10.1186/s12884-016-0987-8

Blair, A., Ritz, B., Wesseling, C., & Freeman, L. B. (2015). Pesticides and human health. *Occupational and Environmental Medicine, 72*(2), 81–82. doi:10.1136/oemed-2014-102454 PMID:25540410

Canella, D. S., Levy, R. B., & Bortoletto Martins, A. P. (2014). Ultra-processed foods and obesity in Brazilian households (2008-2009). *PLoS One, 9*(3), e92752. doi:10.1371/journal.pone.0092752 PMID:24667658

Castro, I., Waclawovsky, G., & Marcadenti, A. (2015). Nutrition and Physical Activity on Hypertension: Implication of Current Evidence and Guidelines. *Current Hypertension Reviews, 11*(2), 91–99. doi:10.2174/1573402111666150429170302 PMID:25921545

Chantal, J., & Serge, H. (2017). Development of a new front-side food labeling system in France: a five-color scale Nutri-Score. *Public Health Panorama, 3*(4), 537–820.

Di, C. A., & Portincasa, P. (2019). Diet and Contaminants: Driving the Rise to Obesity Epidemics? *Current Medicinal Chemistry, 26*(19), 3471–3482. doi:10.2174/0929867324666170518095736 PMID:28521687

WHO. (2003). *Diet, nutrition and prevention of chronic diseases. Report of the joint WHO/FAO expert consultation.* Geneva: World Health Organization. http://www.who.int/nutrition/publications/obesity/ WHO_TRS_916/ru/

Du, H., Guo, Y., Bennett, D. A., Bragg, F., Bian, Z., Chadni, M., Yu, C., Chen, Y., Tan, Y., Millwood, I. Y., Gan, W., Yang, L., Yao, P., Luo, G., Li, J., Qin, Y., Lv, J., Lin, X., Key, T., ... Chen, Z. (2020). China Kadoorie Biobank collaborative group. Red meat, poultry and fish consumption and risk of diabetes: A 9 year prospective cohort study of the China KadoorieBiobank. *Diabetologia, 63*(4), 767–779. doi:10.1007/s00125-020-05091-x

Ferlay, J., Soerjomataram, I., & Dikshit, R. (2015). Cancer incidence and mortality worldwide: Sources, methods, and key patterns in GLOBOCAN 2012. *International Journal of Cancer, 136*, E359–E386. doi:10.1002/ijc.29210pmid:25220842

Filip, L., Miere, D. &Indrei, L.L. (2004). Genetically modified foods. Advantages and humanhealthrisks. *Rev Med Chir Soc Med Nat Iasi, 108*(4), 838–42.

Forastiere, F., & Faggiano, F. (2015). Pesticides, food and health. *Epidemiologia e Prevenzione, 39*(5-6), 305–307. PMID:26554679

GBD. (2016). Risk Factors Collaborators. Global, regional, and national comparative risk assessment of 84 behavioural, environmental and occupational, and metabolic risks or clusters of risks, 1990–2016: A systematic analysis for the Global Burden of Disease Study 2016. *Lancet, 390,* 1345–422.

WHO. (2013). *Global action plan for the prevention and control of noncommunicable diseases. Action Plan for 2013–2020.* World Health Organization.

WHO. (2009). *Global health risks: Mortality and disease burden due to some key risk factors.* World Health Organization.

WHO. (2004). Global strategy on nutrition, physical activity and health. Geneva: World Health Organization.

Hawkes, C. (2010). The influence of trade liberalisation on global dietary change: The case of vegetable oils, meat and highly processed foods. In C. Hawkes, C. Blouin, S. Henson, N. Drager, & L. Dubé (Eds.), *Trade, Food, Diet and Health: Perspectives and Policy Options* (pp. 35–59). Wiley Blackwell.

Heinicke, V., & Halle, M. (2020). Lifestyle intervention in the primary prevention of cardiovascular diseases. *Herz, 45*(1), 30–38. Advance online publication. doi:10.100700059-019-04886-y PMID:31993680

Hercberg, S. (2014). *Propositions pour un nouvel élan de la politique nutritionnellefrançaise de santé publique dans le cadre de la stratégienationale de santé. 1ère partie: mesuresconcernant la préventionnutritionnelle.* Ministère des Solidarités et de la Santé.

Hughes, R. (2004). Competencies for effective public health nutrition practice: A developing consensus. *Public Health Nutrition, 7*(5), 683–691. doi:10.1079/PHN2003574 PMID:15251059

Ji, C., Song, Q., Chen, Y., Zhou, Z., Wang, P., Liu, J., Sun, Z., & Zhao, M. (2020). The potential endocrine disruption of pesticide transformation products (TPs): The blind spot of pesticide risk assessment. *Environment International, 137*, 105490. doi:10.1016/j.envint.2020.105490

Julf & Hemmingsson, E. (2015). Trends in ultra-processed food consumption and obesity in Sweden between 1960 and 2010. *Public Health Nutrition, 18*, 3096–3107. doi:10.1017/S1368980015000506

Lachat, C., Van Camp, J., De Henauw, S., Matthys, C., Larondelle, Y., & Remaut-De Winter, A. M. (2005). A concise overview of national nutrition action plans in the European Union Member States. *Public Health Nutrition*, *8*(3), 266–274. doi:10.1079/PHN2004691 PMID:15918923

Latino-Martel, P., Cotte, V., & Druesne-Pecollo, N. (2016). Alcoholic beverages, obesity, physical activity, and other nutritional factors, as well as cancer risk: an overview of the evidence. *Crit Rev Oncol Hematol, 99,* 308–23. doi: .critrevonc.2016.01.002 doi:10.1016/j

Leonardi, A., Cofini, M., Rigante, D., Lucchetti, L., Cipolla, C., Penta, L., & Esposito, S. (2017). The Effect of Bisphenol A on Puberty: A Critical Review of the Medical Literature. *Int J Environ Public Health, 14*(9). . doi:10.3390/ijerph14091044

Lim, S. S., Vos, T., Flaxman, A. D., Danaei, G., Shibuya, K., Adair-Rohani, H., AlMazroa, M. A., Amann, M., Anderson, H. R., Andrews, K. G., Aryee, M., Atkinson, C., Bacchus, L. J., Bahalim, A. N., Balakrishnan, K., Balmes, J., Barker-Collo, S., Baxter, A., Bell, M. L., ... Ezzati, M. (2012). A comparative risk assessment of burden of disease and injury attributable to 67 risk factors and risk factor clusters in 21 regions, 1990–2010: A systematic analysis for the Global Burden of Disease Study 2010. *Lancet*, *380*(9859), 2224–2260. doi:10.1016/S0140-6736(12)61766-8 PMID:23245609

Liu, P. J., Wisdom, J., Roberto, C. A., Liu, L. J., & Ubel, P. A. (2014). Using behavioral economics to design more effective food policies to address obesity. *Applied Economic Perspectives and Policy*, *36*(1), 6–24. doi:10.1093/aepp/ppt027

Luzada, M., Baraldi, L.G., & Steele, E.M. (2015). Consumption of ultra-processed foods and obesity in Brazilian adolescents and adults. *Prev Med, 81,* 9–15. doi: .ypmed.2015.07.018 doi:10.1016/j

Masharani, U., Sherchan, P., Schloetter, M., Stratford, S., Xiao, A., Nolte Kennedy, M., & Frassetto, L. (2015). Metabolic and physiologic effects from consuming a hunter-gatherer (Paleolithic)-type diet in type 2 diabetes. *Eur J Clin Nutr.*, *69*(8), 944–8. doi:10.1038/ejcn.2015.39

Monteiro, C.A., Moubarac, J.C., Cannon, G., Ng, S.W., & Popkin, B. (2013). Ultra-processed products are becoming dominant in the global food system. *Obes Rev., 14*(S2), 21–8.

Monteiro, C. A., & Cannon, G. (2012). The impact of transnational 'Big Food' companies on the South: A view from Brazil. *PLoS Medicine*, *9*(7), e1001252. doi:10.1371/journal.pmed.1001252 PMID:22802732

Muncke, J. (2011). Endocrine disrupting chemicals and other substances of concern in food contact materials: An updated review of exposure, effect and risk assessment. *The Journal of Steroid Biochemistry and Molecular Biology*, *127*(1-2), 118–127. doi:10.1016/j.jsbmb.2010.10.004 PMID:21073950

Nava-Castro, K.E., Ramírez-Nieto, R., Méndez-García, L.A., Girón-Pérez, M.I., Segovia-Mendoza, M., Navidad-Murrieta, M.S., & Morales Montor, J. (2019). Environmental Pollution as a Risk Factor in Testicular Tumour Development: Focus on the Interaction between Bisphenol A and the Associated Immune Response. *Int J Environ Res Public Health*, *16*(21). doi:10.3390/ijerph16214113

WHO. (2000). *Obesity: Preventing and managing the global epidemic. Report of a WHO consultation*. World Health Organization.

Panel on Contaminants in the Food Chain. (2015). Acrylamide in food. *EFSA Journal*, *13*. . doi:10.2903/j.efsa.2015.4104

Rayner, G., Hawkes, C., Lang, T., & Bello, W. (2006). Trade liberalization and the diet transition: A public health response. *Health Promotion International*, *21*(suppl_1), 67–74. doi:10.1093/heapro/dal053 PMID:17307959

Promoting sustainable consumption – good practices in OECD countries. (2008). OECD Publishing.

Rochester, J.R. (2013). Bisphenol A and human health: a review of the literature. *Reprod Toxicol.*, *42,* 132-55. doi: .reprotox.2013.08.008 doi:10.1016/j

Kedik, S. A., Yartsev, E. I., & Gultyaeva, N. V. (2006). Spirulina – food of the XXI century. Academic Press.

Shephard, G.S. (n.d.). Impact of mycotoxins on human health in developing countries. *Food AdditContam Part A Chem Anal Control Expo Risk Assess*, *25*(2), 146–51. doi: 10.1080/02652030701567442

Serra-Majem, L. (2009). Moving forward in public health nutrition. Introduction. *NutrRev. The I World Congress of Public Health Nutrition*, *67,* S2–S6.

Souza, G.D.S., Costa, L.C.A.D., Maciel, A.C., Reis, F.D.V., & Pamplona, Y.A.P. (2017). Presence of pesticides in atmosphere and risk to human health: a discussion for the Environmental Surveillance. *CienSaude Colet., 22*(10), 3269–3280. doi: 10.1590/1413-812320172210.18342017

Thibault, F., Bernard, S., Laury, S., Emmanuelle, K.G., Benjamin, A., Caroline, M., Mélanie, D., Philippine, F., Paule, L.M., Marie, B., Serge, H., Céline, L., Carlos, A. M., Chantal, J., & Mathilde, T. (2018). Consumption of ultra-processed foods and cancer risk: results from NutriNet-Santé prospective cohort. *The BMJ, 360.* doi:10.1136/bmj.k322

Tsatsakis, A.M., Nawaz, M.A., Tutelyan, V.A., Golokhvast, K.S., Kalantzi, O.I., Chung, D.H., Kang, S.J., Coleman, M.D., Tyshko, N., Yang, S.H., & Chung, G. (2017). Impact on environment, ecosystem, diversity and health from culturing and using GMOs as feed and food. *Food Chem Toxicol., 107*(A), 108–121. doi: 10.1016/j.fct.2017.06.033

World Cancer Research Fund International/American Institute for Cancer Research. (2017). *Cancer preventability estimates for diet, nutrition, body fatness, and physical activity*. wcrf.org/cancer-preventability-estimates

Chapter 3

The State of Children's Health in Light of Malnutrition From Birth to Maturity and Forecasting the Quality of Their Health in the Future

ABSTRACT

This chapter is devoted to the impact of nutrition on the development and growth of nutritionally dependent diseases in children and the further impact on health protection. The role of mother nutrition in child health is noted. Statistical data on the growth of pathological conditions in newborns, which can be triggered by nutritional factors, are presented. The dependence of a child's health on natural feeding of newborns has been substantiated. On the example of statistical data on the state of health of children from birth to puberty, the dynamics of nutritionally dependent diseases over 11 years has been analyzed. The factors leading to the development of some of the nutritionally dependent diseases and the tendencies of the development of diseases are noted. The purpose of the chapter is to substantiate the role of adequate children's nutrition from birth to puberty in the formation of health for many years.

DOI: 10.4018/978-1-7998-8197-1.ch003

INTRODUCTION OF THE CHAPTER

The child's health forms in the womb and depends on many factors, the most important of which is nutritional. Interest in the importance of early life events in lifelong metabolic regulation is increasing every year in connection with new scientific data substantiated by evidence-based medicine. In the early 1990s Hales and Barker (1992), based on convincing epidemiological data, showed a strong association between suboptimal fetal and neonatal nutrition and a number of chronic metabolic conditions at a later age, including cardiovascular disease, hypertension, and diabetes.

The concept of perinatal programming for obesity and diabetes was then extended to other eating disorders, including maternal and / or postnatal overnutrition. It has been suggested that changes in the perinatal environment may affect the structure and function of key metabolically significant organs such as the pancreas, liver and adipose tissue. There is also a growing understanding that programming the development of neural systems involved in the energy balance in the perinatal environment is a potential cause of obesity and diabetes. An important component of this nervous system are neurons in the hypothalamus (Bouret S.G. 2017, Dearden L.et al., 2018).

The consequence of malnutrition from birth is an increase in the number of nutritional diseases in children.

Further neglect of food culture can adversely affect the health. The lack of native essential components in sufficient quantities can lead to the development of microelementosis, changes in the body's immune response and, as a consequence, a decrease in resistance to viral and bacterial diseases.

The maximum transition to refined food, which is all fast food, contributes to the progression of nutritionally dependent diseases in adulthood.

MAIN FOCUS OF THE CHAPTER 3

3.1 Influence of Maternal Nutrition on Child Health and Development

The quality and diet of the mother, the variety of food components, the quality of the food and the amount of digestible food substances determine the course of pregnancy and the general health of the baby, as well as the body's immune reserve and nutritional preferences in the future. The uniformity of nutrition, the deficiency of essential substances, the cumulative effect of contaminants can lead to a decrease in the resistance of the mother's body and, as a result, changes in the health of the fetus, up to the teratogenic effect (Mother's healthy diet: a better start to life. World Health Organization. 2016).

Along with other causes, malnutrition during pregnancy can lead to grave consequences - miscarriage, premature birth, childbirth with various intrauterine defects, lag in physical and mental development.

During pregnancy, several risk factors can harm a woman's health. For example, pregnancy in adolescence places a double burden on the body in terms of nutrition. While a teenage girl needs to meet the growing demands for nutrients for her own physical growth, the additional needs of the fetus threaten the nutritional well-being of both mother and a child.

Pregnancy of adolescents requires special attention, as young mothers and their children are at greater risk of food insecurity and social and health problems. Another important aspect of pregnancy is the need for an adequate intake of micronutrients, especially iron, folate and iodine (Robertson A. et al., 2004). Micronutrient deficiency can develop during prenatal development and largely depends on the nutritional status of the mother. Deficiency of certain macro- and micronutrients in the prenatal period affects not only the formation of the fetus, but also negatively affects the future state of the child's health and is a risk factor for the development of a number of diseases in adulthood. Thus, the nutrition of a woman during pregnancy should ensure the correct course, a favorable outcome of pregnancy, maintain an optimal state of a woman health, and contribute to the normal formation, growth and development of the fetus.

In recent years, much attention has been paid to the sufficient intake of zinc during pregnancy. Experimental studies have revealed a pronounced teratogenic effect: abnormalities in the brain development, cleft of the upper jaw, anomalies in the development of the heart, lungs, and urogenital system. Zinc deficiency affects the development of the embryo and the fetus through several mechanisms, including a decrease in cell proliferation, a decrease in protein synthesis, an increase in the activity of free radical processes and cell apoptosis. Zinc deficiency in a pregnant woman is accompanied by a decrease in the immune response, increased incidence. It increases the frequency of pregnancy complications and relapses of chronic diseases.

Iron deficiency in a pregnant woman increases the risk of premature birth. It deficiency negatively affects the immune response and increases the risk of infection; at the same time, the activity of phagocytes and neutrophils decreases and the bactericidal effect of K cells decreases. But these are just some of the factors affecting the formation of the fetus health, and subsequently, the child. Table 1 shows the types of deficient conditions of nutritional substances and possible consequences for the fetus (FAO/WHO, 2016-2018; The national program for optimizing the feeding of infants in the Russian Federation. Moscow: Union of Pediatricians of Russia, 2010).

Table 1. Types of deficient conditions of nutritional substances and possible consequences for the fetus

Deficientconditions	Developmental anomalies of the fetus
Protein and energy failure	Intra-uterine hypotrophy, Brain microplasia
Deficiency of long chain polyunsaturated fatty acids, violation of the ratio of ω-3 and ω -6 PUFA	Impaired development of brain tissue and retina. Growth retardation, skin disorders, reproductive disorders, fatty degeneration of the liver, impaired cognitive function
Folicaciddeficiency	Impaired development of the neural tube (cerebral hernia, anencephaly, etc.), slow growth of enterocytes
Vitamin A deficiency	Congenitalanomaly
Iodinedeficiency	Neurodevelopmental disorder (in severe forms - congenital cretinism)
Irondeficiency	Decreased cytotoxic activity of phagocytes and proliferation of T1 lymphocytes
Zincdeficiency	Decreased proliferation and capacity of T1 lymphocytes
Seleniumdeficiency	Decreased antibody production, cytokine synthesis, lymphocyte proliferation
Copperdeficiency	Decreased phagocytic activity, antibody production, T lymphocyte proliferation and B cell number

Micronutrient deficiency causes intrauterine development and growth retardation, which can persist for the first two years and permanently impair cognitive abilities and other functions in later life. Even if micronutrient deficiency is corrected in subsequent periods of childhood, correction of early consequences is impossible, that is why, the problem of growth retardation can transmit from one generation to another (Robertson A. et al., 2004).

The sufficient intake of folic acid is of great importance for the prevention of such birth defects, as cleft spine. The additional administration of folic acid preparations before and during the first weeks of pregnancy prevents up to 75% of neural tube defects in the developing fetus (Molloy, A.M. & Scott, J., 2001). More and more data indicate that folic acid in the mother's body can reduce the likelihood of developing cardiovascular disease in the subsequent life of the child (Molloy, A.M. & Scott, J., 2001). The most vulnerable stages of life belong to the first weeks of pregnancy, when a woman may not know about her state. Therefore, the provision of a woman with folate should always be sufficient, since many pregnancies are not planned in advance.

Recently, the number of children born with body weight below normal has increased. Table 2 shows the data for the Russian Federation. They evidenced a sharp increase in the birth of children with insufficient body weight. The increase is especially noticeable in the group with a body weight at birth of 500 - 999 g: in the period from 2010 (over 6 years) an increase is more than 60%. Over 11 years, the birth of children in this group increased by 110%.

The number of children with body weight at birth of 1000-1499 g continues to increase. A slight decrease was noticed only in 2016 and is most likely associated with beneficial medical technologies in the labor management and obstetric care.

It should be noted that the number of children with body weight below the norm in relation to all newborns by years is approximately the same in percentage terms and amounts to 5.8 - 6.2, but the growth in groups is more significant and is expressed in the increase in newborns with a weight of 500 - 999 and 1000-1499 g, respectively.

This trend is global. Low birth weight (less than 2500 g) is the result of either premature birth or intrauterine growth retardation, which may be associated with poor nutritional status of the mother.

Table 2. Data on newborns by body weight at birth in the Russian Federation

	2005	2010		2013		2014		2015		2016	
	Number	Number	% of 2005	Number	% of 2005	Number	% of 2005	Number	% of 2005	Number	% of 2005
Bornalive											
In total	1418962	1762782	24.2	1866442	31.5	1910494	34.6	1909420	34.6	1857301	30.9
of which with birth weight, g:											
500 - 999	3163	4724	49.4	6873	117.3	6740	113.1	6650	110.3	6645	110.1
1000-1499	6767	9633	42.4	12218	80.6	12355	82.6	12599	86.2	11868	75.4
1500-1999	18726	22653	21.0	24726	32.04	24758	32.2	24435	30.5	23637	26.3
2000-2499	55949	65714	17.5	71121	27.12	71127	27.1	70888	26.7	69111	23.5

* Health care in Russia. (2017): statistical book / Rosstat. Moscow

Table 3. The incidence of children of the first year of life by major classes and groups of diseases in the Russian Federation

Diseases	2005	2010		2013		2014		2015		2016	
	Number	Number	% of 2005	Number	% of 2005	Number	% of 2005	Number	% of 2005	Number	% of 2005
Registered diseases, in total among them:	3625.4	4419.3	21.9	4449.7	22.7	4421.2	22.0	4399.5	21.4	4761.6	31.3
diseases of the central nervous system among them:	227.2	331.2	45.8	372.9	64.1	385.2	69.5	393.6	73.2	434.4	91.2
infantilecerebralparalysis	1.2	1.8	50.0	1.1	-8.3	1.3	8.3	1.2	0.00	1.5	25.0
diseases of the eye and its appendages	104.4	150.2	43.9	163.9	57.0	171.0	63.8	173.6	66.3	195.5	87.3
diseases of the respiratory system	1603.1	2004.6	25.1	2080.8	29.8	2088.2	30.3	2090.2	30.4	2299.2	43.4
diseases of the digestive system	190.7	250.0	31.1	244.7	28.3	240.2	26.0	253.9	33.1	255.3	33.9
diseases of the genitourinary system	59.3	84.3	42.2	88.3	48.9	87.8	48.1	89.4	50.7	100.3	69.1
congenital malformations, deformations and chromosomal abnormalities	84.7	125.3	47.9	124.7	47.3	130.0	53.5	130.5	54.1	150.5	77.7

* Health care in Russia. (2017): statistical book / Rosstat. Moscow

Surprisingly, among the countries of Central and Eastern Europe, only Macedonia is characterized by a significantly higher prevalence than Western European countries. In the European Union, the share of low birth-weight babies is relatively unchanged over the past 20 years: a decrease was observed only in the countries of Northern Europe. In post-Soviet countries (USSR), in contrast, the number of low birth-weight babies has increased.

It should be noted that the differences in the reporting system and in the determination of low birth weight in different countries can vary significantly. That is why statistics is not always indicative and reflects the true picture. For example, in some countries, children born with a weight of less than 1000 g are not included in reporting because of the too high risk of death. The high prevalence of low birth weight can be one of the signs of nutritional deficiency in the mother's diet. In a study of socially determined differences in the outcome of childbirth conducted in the Czech Republic, it was noted that the average body weight at birth from 1989 to 1991 decreased and then increased again. From 1989 to 1996, the difference in average birth weight between children born of mothers with primary education and mothers with higher education increased by about 40%. This means that the relatively worse outcome of childbirth in less educated mothers may be related to their nutrition before and during pregnancy, and that socially determined differences in the outcome of childbirth in the European Region could increase. Smaller reserves of nutrients in the body not only cause physical defects in young children, but can also hinder the development of their immune system and brain.

Recent studies by a number of scientists (Thayer Z.M. et al., 2020) show that birth weight heavily depends on the mother's usual diet before pregnancy to a greater extent than diet during pregnancy and emphasize the need for solutions aimed at optimizing future, intergenerational health results.

An increase in the share of children born with insufficient body weight and an increase in the percentage of female infertility (mentioned in the Chapter 2) may indirectly indicate a negative trend in the deterioration of the health of the population as a whole and a significant effect of the nutritional factor.

The results of scientific studies conducted in recent years show that the nutrition of a child affects not only his growth, development and health status. It became apparent that nutrition in the first year of life "programs" metabolism in such a way that certain malnutrition can increase the risk of developing a number of diseases, such as allergic diseases, obesity, metabolic syndrome, osteoporosis and some others. It should be noted that all these diseases are currently called the "epidemic" of civilization, as evidenced by the increasing frequency of cases, the severity of the course and the many adverse consequences.

Fetal programming, or the fetal germination hypothesis, suggests that changes in the nutrition and endocrine status of the fetus lead to adaptations in the development process that predispose the individual to CVD and metabolic and endocrine disorders in later life (Godfrey K.M. & Barker D.J., 2001).

Factors that need to be considered during the period of intrauterine development are the following:

- intrauterine growth retardation, which can lead to low birth weight;
- premature birth of a fetus with normal physical development for fetal age;
- factors transmitted from one generation to another (Robertson A. et al., 2004).

The existence of metabolic programming was proven initially for babies born prematurely or with intrauterine malnutrition. Clinical studies of large cohorts of children have revealed that low birth weight contributes to early arterial hypertension, as well as the development of insulin resistance at the age of 20-30 years.

New scientific studies have allowed us to demonstrate that the same complications are observed in children who have a high growth rate in the first year of life.

Along with this, hypotrophy in children of the first year of life is also distinguished, which manifests itself in the form of protein-energy malnutrition with a significant decrease in immunity and other criteria.

The following causes of malnutrition in children are distinguished:

- congenital malformations (cardiovascular and bronchopulmonary systems, gastrointestinal tract, liver, urinary system);
- congenital or acquired CNS involvement (cerebral ischemia, intracranial hemorrhage, etc.);
  malabsorption syndrome (celiac disease, food allergy, cystic fibrosis, short bowel syndrome, etc.);
- congenital immunodeficiency conditions;
- some endocrine diseases, etc. (panhypopituitarism, Addison's disease, dysfunctions of the thyroid gland, etc.);
- hereditary metabolic disorders;
- long-term intoxication in chronic infectious diseases (tuberculosis, brucellosis, etc.) and purulent processes (abscesses, suppurating bronchiectasis, osteomyelitis);
- severe acute infectious diseases, accompanied by intoxication, vomiting, loose stool;
- severe non-infectious diseases (malignant tumors, severe heart disease, respiratory failure, etc.);
- severe thermal and combined injury;
- alimentary reasons (quantitative and qualitative malnutrition during natural and artificial feeding, improper introduction of complementary foods, defects in care).

Analyzing the incidence of children in the first year of life according to the main classes and groups of diseases caused by alimentary factors (Table 3), an increase in the number of diseases over the past 11 years should be noted. At the same time, the greatest sharp in growth is observed among diseases of the nervous system (by 91.2%). This may be due to both the insufficient presence of polyunsaturated fatty acids not subjected to oxidative processes, and an insufficient amount of vitamins such as thiamine, pyridoxine and antioxidants.

Folacin deficiency during pregnancy can have a teratogenic effect, as well as lead to impaired mental development of newborns (Pokrovsky A.A., 1981).

An increase in diseases of the eye and its adnexa is also marked over 11 analyzed years (by 87.3%). This may be due to a deficiency of vitamin A (retinol). At present, vitamin A deficiency is one of the most serious diseases of nutritional deficiency, affecting primarily children and is one of the most important causes of eye diseases, including acquired blindness.

The cause of vitamin A deficiency is a deficiency of animal foods rich in vitamin A, as well as protein deficiency, accompanied by a violation of the synthesis of retinol-binding protein and the processes of retinol absorption and transport. These processes are closely associated with impaired absorption of lipids in the intestine, which depends on the state of the intestinal mucosa and the hepatobiliary system (chronic enteritis, etc.). Literacy in the diet formulation of pregnant women and nursing mothers and the inclusion of animal products rich in vitamin A and plant foods rich in carotenoids (beta carotene, lutein, zeaxanthin, lycopene) in the diet also play a significant role. An important role is played by the provision of the diet of pregnant and subsequently lactating women with the multivitamin complexes for at least 6 months a year ("The National Program for Optimization of Provision with Vitamins and Minerals of Children in Russia", 2017).

3.2. The Importance of Breast Feeding

The best food for the baby in the first months of life is mother's milk, which corresponds to the peculiarities of its digestive system and metabolism and ensures the adequate development of the baby's body at the rational nutrition of a nursing woman. All nutrients of breast milk are easily absorbed, because their composition and ratio correspond to the functional capabilities of the gastrointestinal tract of the infant, as well as due to the presence of enzymes (amylase, lipase, phosphatase, protease, etc.) and transport proteins. Breast milk is a source of hormones and various growth factors (epidermal, insulin-like, etc.), which play a crucial role in the regulation of appetite, metabolism, growth and differentiation of baby's tissues and organs.

Due to antibodies, immune complexes, active white blood cells, lysozyme, macrophages, secretory immunoglobulin A, lactoferrin and other biologically active substances, breast milk increases the protective functions of the child's body. Oligosaccharides, as well as low levels of protein and contribute to the growth of healthy intestinal microflora.

In recent years, due to the decoding of the genome of microorganisms that inhabit the body's natural niches, it has become apparent that breast milk is unsterile, it contains more than 700 types of microbiotic strains from opportunistic (Staphylococcus aureus, Proteus) to indigenous (normal) microflora: bifidobacteria and lactobacilli that determine the formation of immunity, programming the growth and development of the child, the formation of his mental and even social functions.

Therefore, children who are naturally fed are much less likely to suffer from infectious diseases, and develop more stable post-vaccination immunity. The protective properties of breast milk are not limited to anti-infective protection. Breastfeeding reduces the risk of development of such diseases as atherosclerosis, hypertension, diabetes mellitus, obesity, leukemia, etc.

Breast feeding has a beneficial effect on the development of the central nervous system of the child and its mental status. The unity of mother and baby during breastfeeding has a deep mutual emotional effect. It was noted that children who were breast-fed are characterized by harmonious physical development, they are calmer, more balanced, and friendly compared to children who were formula-fed. They become attentive and caring parents.

Metabolic factors that act for a limited and sensitive period of pre- and postnatal development can have long-term effects on health and the risk of disease development at a later age up to an old age, including the later risk of obesity, which is called early metabolic programming for long-term health.

B. Koletzko et al. (KoletzkoB. et al., 2010) indicate that the risk of obesity in school age was reduced with early breastfeeding compared with formula feeding. According to the assumptions of this group of scientists, breastfeeding protects against late obesity, reducing high weight gain in infancy.

One reason is the lower protein content in breast milk compared to conventional infant formulas (the "early protein hypothesis"). This hypothesis was tested in the European project on childhood obesity - a double-blind, randomized clinical trial, which involved 1678 infants in five countries (Belgium, Germany, Italy, Poland, Spain).

One randomized group of healthy babies born on time in the first year of life was fed with formula with a higher protein content, the other group of babies was fed with lower protein formula, respectively. Subsequent data, obtained at the age of 2 years, suggests that the low-protein formula normalizes early growth relative to the breastfeeding comparison group and the current WHO growth standard, which can provide significant long-term protection against subsequent obesity.

Scientists have concluded that infant feeding practices have high potential for long-term health effects. The results should stimulate the review of recommendations and policies regarding the composition of infant formulas (Koletzko B. et al., 2010).

According to some reports, breast-fed infants have a higher intellectual development rate, which is probably partly due to the presence of long-chain polyunsaturated fatty acids in breast milk, which are necessary for the development of brain and retinal cells.

In the blood of breast-fed infants, the amount of long-chain polyunsaturated fatty acids is significantly higher compared with formula-fed infants. The protein of breast milk consists mainly of whey proteins (70–80%), which contain essential amino acids in a ratio optimal for the child, and casein (20–30%). Protein fractions of breast milk are divided into metabolizable (food) and non-metabolizable proteins (immunoglobulins, lactoferrin, lysozyme, etc.), which are 70–75 and 25–30%, respectively. In contrast to cow's milk, breast milk contains a large amount of alpha-lactalbumin (25–35%), which is rich in essential and conditionally essential amino acids (tryptophan, cysteine). Alpha-lactalbumin promotes the growth of bifidobacteria, the absorption of calcium and zinc from the child's gastrointestinal tract, the normalization of the child's mood, the formation of the wakefulness and sleep patterns.

The level of breast milk protein of not more than 1.2 per 100 ml in the first half of life has the programming metabolic force, as well as 1.3 per 100 ml of breast milk in the second half of life, despite the high growth rates of growth and weight of the baby.

Adequate metabolic programming is to optimize the utilization of amino acids, the absence of "unused" amino acids that stimulate the excessive formation of an insulin-like growth factor, its circulation in the blood and stimulation of insulin release, and, as a result, enhanced adipogenesis (excessive formation of fat cells) and weight gain. Thus, the protein level in breast milk programs the normal weight of the baby in the future.

The composition of breast milk contains nucleotides, which account for about 20% of all non-protein nitrogen. Nucleotides are the initial components for the formation of ribonucleic and deoxyribonucleic acids; they play an important role in maintaining the immune response, the growth stimulating and enterocytes differentiation.

The main components of breast milk fat are triglycerides, phospholipids, fatty acids, and sterols. Its fatty acid composition is characterized by a relatively high content of essential polyunsaturated fatty acids (PUFAs), the concentration of which in breast milk is 12-15 times higher than in cow's milk.

PUFAs are precursors of arachidonic, eicosapentaenoic and docosahexaenoic fatty acids, which are an important component of cell membranes; various classes of prostaglandins, leukotrienes and thromboxanes are formed from them; they are also necessary for myelination of nerve fibers and the formation of the retina.

Long-chain polyunsaturated fatty acids (arachidonic and docosahexaenoic) are found in breast milk in a small amount (0.1–0.8 and 0.2–0.9% of the total fatty acid content, respectively), but their level is significantly higher than in cow's milk.

Fats of mother's milk are easier to digest, as they are more emulsified. In addition, breast milk contains lipase, which is involved in the digestion of the fat milk component.

Healthy intestinal microflora has programmatic significance, its species diversity depends on the mother's nutrition not only during pregnancy, but also during breastfeeding, on the method of delivery (cesarean section significantly reduces species diversity), as well as on antimicrobial therapy reprogramming diversity and species composition of intestinal microflora).

It should be noted that if the mother's diet contains in large number of contaminants of anthropogenic nature, the effect of breastfeeding is reduced.

It should be noted that there is no adequate alternative to breastfeeding, despite significant advances in the development and production of baby food for an early age.

3.3. Dynamics of Diseases from Birth to Puberty. Prevention of Nutritionally Dependent Diseases in Children

It is impossible to predict changes in congenital anomalies (malformations), deformations and chromosomal abnormalities depending on one decisive factor, but nutritional factor can play a significant role. The number of contaminants in food increases from year to year; this can lead to increased mutagenic and teratogenic effects. These tables 4 clearly show an increase of 77.7% over the past 11 years in congenital anomalies (malformations), deformations and chromosomal abnormalities.

It should be noted an increased number of diseases of the genitourinary system in children up to a year. Compared to 2005, to 2010 it increased by 42%, and to 2016 by 69.1%. The dynamics are obvious. It is difficult to predict the reasons for this growth, but it can be assumed that an increase in nitroso-compounds in the diet promotes the synthesis of nitrosamines in the human body (pregnant woman and fetus). Nitroso-compounds have carcinogenic, mutagenic, teratogenic and embryotoxic properties. The target of the action of large doses of nitrates are hepatocyte nuclei and nucleic acid metabolism, which explains mainly the embryotoxic effect.

In global food products, nitroso-compounds are found not only in sausages and delicatessen meat products, where they are a color stabilizer, but they are also used as preservatives in cheeses, salted fish and other products. Another source of nitroso-compounds is fresh and green vegetables grown with intensive cultivation using pesticides and fertilizers.

Respiratory diseases in children under one year of age increased by 43.4% over 11 years, and digestive diseases by 33.9%, respectively.

Statistical data on registered diseases do not show some "features" of nutritional factors, that is why food intolerance and food allergy are increasingly evident factors. And although the first unusual reactions to food have been known since the time of Hippocrates (460-370 BC), they are most widespread in the age of food globalization.

In recent decades, there is a lot of information on the growth of allergic, autoimmune diseases and other conditions associated with impaired immune system function. The development of allergic sensitization in infancy indicates the inability of the body to develop tolerance to the food antigen (Muraro A et al., 2014, Nazamova-Baranova L.S. 2011). In the chain of incorrect approaches to feeding a child, the leading and strengthening role belongs to the early introduction of cow's milk into the baby's diet, first in the form of an adapted standard formula, then in the form of whole cow's or goat's milk.

Food allergy (FA) is a state of hypersensitivity to food products, which is based on immunological mechanisms, the leading ones are IgE-mediated reactions. Food allergy is a pathological reaction based on immune mechanisms (specific IgE-mediated reactions, a cellular immune response (not IgE-mediated), or thier combination – mixed-type reactions)[1].

When coding food allergy, codes of the main symptoms detected in patients are used. The highest prevalence of food allergy is observed in children of the first two years of life and amounts to 6–8%. Currently, more than 160 food allergens that cause IgE-mediated allergic reactions in children have been identified. The list of "Big Eight" products with the highest allergenicity includes cow's milk, eggs, fish, wheat, peanuts, soy, crustaceans (shrimp, crabs, lobsters and others), nuts (hazel nuts, almonds, walnuts).

Widespread allergens also include cocoa and chocolate, citrus fruits, strawberry, honey, meat of animals and birds, caviar and other seafood, cereals.

In infants, one of the first allergens that cause the development of food allergy is cow's milk. It contains 15–20 antigens, of which casein, γ-lactoglobulin, α-lactalbumin, and bovine serum albumin are the most allergenic. Cow's milk ingredients (casein, whey proteins) can be present in many foods. It is known that allergy to cow milk proteins is marked in 0.5–1.5% of breast-fed infants, and up to 2–5% of formula-fed infants. In patients with atopic dermatitis, cow milk protein allergy is diagnosed in 70–85% of cases.

Allergic reactions associated with the intake of food products can manifest themselves with many symptoms due to the involvement of various organs and systems in the pathological process. For young children, the most characteristic are skin manifestations of allergy (atopic dermatitis, frank intertrigo and prickly heat, Quincke's disease, urticaria), as well as various gastrointestinal disorders (profuse regurgitation and vomiting, colic, a change in the nature of the stool - diarrhea or constipation).

Rarer manifestations of food allergy include respiratory disorders (rhinitis, shortness of breath, apnea) and severe general anaphylactic reactions (Ferreira C. & Seidman E. 2007).

An analysis of the incidence of children aged 0-14 years by main classes and groups of diseases (diagnosed for the first time in life) and having nutritional nature (Table 4) showed a significant increase in diabetes mellitus by 96.3%. It is also registered a high increase in neoplasms (62.52%), including those registered for the first time (Table 5), and in obesity (64.52%).

The increase in these diseases over the previous 11 years may be due to the quality of food. Increasingly, children and adolescents prefer snacks in fast food enterprises for nutritious meals. The number of confectionery products with easily digestible sugars often exceeds the daily requirement by several times. Diet is rarely distinguished by a variety of essential substances necessary to maintain the health of the younger generation.

Table 4. Incidence of children aged 0-14 years by main classes and groups of diseases with nutritional nature in the Russian Federation

Diseases	2005	2010		2013		2014		2015		2016	
	Number, thousand persons	Number, thousand persons	% of 2005	Number, thousand persons	% of 2005	Number, thousand persons	% of 2005	Number, thousand persons	% of 2005	Number, thousand persons	% of 2005
All diseases	36837.4	40903.5	11.04	43155.1	17.15	44157.6	19.87	43843.0	19.02	44831.9	21.70
Among themneoplasms	73.1	96.2	31.60	110.3	50.89	115.4	57.87	118.0	61.42	118.8	62.52
Among themdiabetesmellitus	2.7	3.1	14.81	3.9	44.44	4.5	66.67	4.7	74.07	5.3	96.30
obesity	55.8	69.2	24.01	83.7	50.00	87.0	55.91	93.2	67.03	91.8	64.52
Diseases of the respiratory system	21780.3	25507.7	17.11	27341.4	25.53	28176.3	29.37	28578.3	31.21	29656.8	36.16
Congenital anomalies, deformations and chromosomal abnormalities	199.0	255.5	28.39	267.4	34.37	277.9	39.65	270.6	35.98	274.8	38.09

(registered diseases in patients with a diagnosis established for the first time in their life)
*Health care in Russia. 2017: statistical book (2017)/ Rosstat. Moscow,

Table 5. The incidence of children aged 0-14 years with malignant neoplasms in the Russian Federation

2005	2010		2013		2014		2015		2016	
Number	Number	% of 2005	Number	% of 2005	Number	% of 2005	Number	% of 2005	Number	% of 2005
Diseases registered in patients with a diagnosis established for the first time in their lives, thousand persons										
2.3	2.7	17.39	2.9	26.09	3.0	30.43	3.2	39.13	3.3	43.48
The number of patients registered in medical organizations (at the end of the year), thousand persons										
11.9	13.9	16.81	15.4	29.41	16.5	38.66	17.5	47.06	19.1	60.50

*Health care in Russia. 2017: statistical book (2017)/ Rosstat. Moscow

Overweight by only 20% increases the likelihood of diabetes by 2-2.5 times, which can be seen from the data in Table 4. And since physical activity is reduced in children during globalization, the consumption of refined food products aggravates the state of health and leads to a sharp increase in nutritional-dependent diseases.

An increase in respiratory diseases is associated indirectly with the nutritional factor, most children and adolescents have reduced body resistance and are unable to resist infection. In addition, some respiratory diseases are caused by allergic reactions, including allergy to food.

As noted above, the number of malignant neoplasms is steadily growing and progressing (Table 5). A growing number of children under 14 years old are diagnosed with these diseases. There is an interesting difference between the number of diseases in patients with a diagnosis established for the first time in their lives and the number of patients registered in medical organizations (at the end of the year).

If in 2010, the share as of 2005 between the number of diseases in patients with a diagnosis established for the first time in their life and the number of patients registered in medical organizations (at the end of the year) was 17.39 and 16.81%, in 2016, the share as of 2005 of the same indicators was 43.48 and 60.50%, respectively. Obviously, the number of patients registered in medical organizations (at the end of the year) is growing steadily.

The number of neoplasms is also growing, and for another age group of 15-17 years, the share over 6 years (from 2010 to 2016) has more than doubled (from 23.3% from 2005 to 2010 to 50.05% 2010 to 2016).

In the age group of 15-17 years, it is noted an increase in diseases of the blood, blood-forming organs and certain disorders involving the immune mechanism from 2005 to 2010 by 23.82%, to 2013 by 45.81% and a slight decrease in 2016 – 38.04% (in relation to 2005).

Anemia is a predominant disease in this group that makes up from 89.2 to 92% of all diseases of the blood, blood-forming organs and certain disorders involving the immune mechanism. Anemia is growing every year. And if the difference between 2010 and 2005 was 28.94%, then by 2013 it had increased to 46.60% in comparison with 2005, then there was a slight decrease, but compared with the growth it was not significant. This fact can be associated with a decrease of easily assimilated iron (heme iron), vitamin C in the diet, which contributes to the iron absorption. In addition, the development of anemia is influenced by anti-nutritional factors, as well as nitrites in food products.

In the age group of 15-17 years, leading diseases are those directly related to the nutritional factor - obesity and type 2 diabetes. In comparison with the age group of 0-14 years, the growth of these diseases is high. From 2005 to 2010, the percentage of obese children increased by 60.37, and by 2016 the increase was 153.60%. The number of patients with diabetes in this age group increased by 98.09% over 11 years (Table 6).

It should be noted the increase in epilepsy diseases and status epilepticus; by 2016, the growth was 30.49%, while by 2010 it was 6.10% compared with 2005.

In 2010, digestive apparatus diseases increased by 19.03%, and 2013 - by 35.64%, then in 2016 there was a significant decrease by 17.13%.

Table 7 shows the distribution of the number of children with disabilities aged 0-17 years by diseases that caused disability. The total number of diseases for which disability has been established is slightly reduced. At the same time, there is a significant increase in a number of tumors and in the dynamics over the years. In 2016, the number of disabled children aged 0-17 years who were diagnosed with a disability due to neoplasms increased by 20, 37%, and due to diabetes mellitus – by 45.07%, respectively.

Table 6. The incidence of children aged 15-17 years according to the main classes, groups and certain diseases (diseases registered in patients with a diagnosis established for the first time in life) in the Russian Federation

Diseases	2005	2010		2013		2014		2015		2016	
	Number	Number	% of 2005	Number	% of 2005	Number	% of 2005	Number	% of 2005	Number	% of 2005
	Per 100 000 children aged 15-17										
All diseases	111451.9	135763.3	21.81	143754.8	28.98	141653.0	27.10	137383.8	23.27	137273.8	23.17
Among themneoplasms	312.9	385.8	23.30	450.2	43.88	474.6	51.68	501.1	60.15	469.5	50.05
Diseases of the blood, blood-forming organs and certain disorders involving the immune mechanism	673.0	833.3	23.82	981.3	45.81	965.0	43.39	907.4	34.83	929.0	38.04
Among themanemia	600.2	773.9	28.94	879.9	46.60	846.2	40.99	840.6	40.05	853.5	42.20
Diabetesmellitus	15.7	16.8	7.01	22.2	41.40	23.6	50.32	27.8	77.07	31.1	98.09
Obesity	289.2	463.8	60.37	582.8	101.52	638.5	120.78	732.8	153.39	733.4	153.60
Epilepsy, statusepilepticus	68.9	73.1	6.10	72.6	5.37	80.8	17.27	83.7	21.48	90.6	31.49
Diseases of the digestive system	5945.3	7076.9	19.03	7885.6	32.64	8064.4	35.64	7315.0	23.04	6964.0	17.13

*Health care in Russia. 2017: statistical book (2017)/ Rosstat. Moscow

Table 7. Distribution of the number of disabled children aged 0-17 years by diseases that caused the onset of disability in the Russian Federation

Diseases	2005	2010		2013		2014		2015		2016	
	Number	Number	% of 2005	Number	% of 2005	Number	% of 2005	Number	% of 2005	Number	% of 2005
Per 100 000 children											
All diseases	198.0	188.4	-4.85	192.2	-2.93	192.7	-2.68	188.5	-4.80	187.1	-5.51
Neoplasms	5.4	5.5	1.85	6.1	12.96	6.3	16.67	6.4	18.52	6.5	20.37
Endocrine diseases, eating disorders, metabolic disorders	11.1	11.4	2.70	12.2	9.91	12.7	14.41	12.8	15.32	13.6	22.52
Among them diabetesmellitus	7.1	7.8	9.86	8.9	25.35	9.4	32.39	9.6	35.21	10.3	45.07
mentalandbehavioraldisorders	46.9	46.5	-0.85	48.2	2.77	48.3	2.99	48.6	3.62	49.1	4.69
diseases of the central nervous system	41.6	42.0	0.96	44.3	6.49	44.8	7.69	45.0	8.17	45.2	8.65
Congenital anomalies, deformations and chromosomal abnormalities	36.4	36.2	-0.55	35.8	-1.65	36.1	-0.82	33.3	-8.52	32.0	-12.09

*Health care in Russia. 2017: statistical book (2017)/ Rosstat. Moscow

A decrease in establishing the status of a disabled child is observed only for congenital anomalies (malformations), deformations and chromosomal abnormalities. Despite the fact that the total number of congenital anomalies (malformations), deformations and chromosomal abnormalities increases over the years, their nature does not always imply a decrease in legal capacity.

Currently, Russia has developed "The National Program for Optimizing the Feeding of Infants in the Russian Federation" (2019) and the "National Program for Optimizing the Provision of Vitamins and Mineral Substances for Children in Russia" (2017). These programs are aimed at the following measures:

- knowledge increment of pediatricians, neonatologists, obstetrician-gynecologists on nutrition of healthy and sick infants, pregnant women and nursing mothers;
- adaptions to the educational programs of medical universities on the organization of child nutrition;
- an increase in breastfeeding by 25-30% in the Russian Federation;
- a 15% reduction in the incidence of nutritional-dependent pathology in young children due to the optimization of their nutrition, including

the use of modern industrial products enriched with vitamin-mineral complexes, pre- and probiotics, nucleotides, long-chain fatty acids;

- improving the health status of infants and young children by reducing the incidence rate and preventing disability through the introduction of modern technologies for the organization of rational and therapeutic nutrition;
- optimization of the supply of pregnant, lactating and children of any age with vitamins and minerals.

This program is given as an example of existing developments, and like many programs it has a significant drawback - there are no recommendations for the consumption of modern food products, many of which are refined and are primarily energy value.

The development of real recommendations for the correct consumption of global food products, based on the research of modern scientists, is necessary and obvious.

CONCLUSION OF THE CHAPTER

Based on statistical data on the increase in morbidity in children and adolescents, as well as an increase in newborns with pathological conditions, it can be noted that the health of an adult is formed from the conception and directly depends on the mother's nutrition and her health state. Lack of essential substances (latent hunger) can provoke fetal failure and a development of certain diseases. Excessive maternal nutrition, especially in fats and digestible carbohydrates, can lead to the tendency to diabetes mellitus and obesity in children.

Anthropogenic contaminants, especially those capable of migrating along the food chain, are a negative factor in the nutrition of children from birth to puberty. The amount of harmful substances increases from year to year, and regulatory authorities often do not have time to regulate new substances. Consequently, there is a high likelihood of uncontrolled foreign substances entering the food of children and adolescents and their impact on health.

The growth of nutritionally dependent diseases in children is becoming threatening. In adulthood, childhood diseases and (or) pathologies will progress and significantly worsen life quality. Therefore, it is important to conduct research on the impact of modern foods on human health in the long term (with regular consumption).

REFERENCES

Barker, D. J. P. (1998). *Mothers, babies and health in later life* (2nd ed.). Churchill-Livingstone.

Bouret, S. G. (2017). Development of Hypothalamic Circuits That Control Food Intake and Energy Balance. In R. B. S. Harris (Ed.), *SourceAppetite and Food Intake: Central Control* (2nd ed.). CRC Press/Taylor &Francis. doi:10.1201/9781315120171-7

Dearden, L., Bouret, S.G., &Ozanne, S.E. (2018). Sex and gender differences in developmental programming of metabolism. *Mol Metab, 15,* 8–19. doi:10.1016/j.molmet.2018.04.007

Ferreira, C., & Seidman, E. (2007). Food allergy: A practical update from the gastroenterological viewpoint. *Jornal de Pediatria, 83*(1), 7–20. doi:10.2223/JPED.1587 PMID:17279290

Godfrey, K. M., & Barker, D. J. (2001). Fetal programming and adult health. *Public Health Nutrition, 4*(2b, 2B), 611–624. doi:10.1079/PHN2001145 PMID:11683554

Koletzko, B., Schiess, S., Brands, B., Haile, G., Demmelmair, H., von Kries, R., & Grote, V. (2010, July). Infant feeding practice and later obesity risk. Indications for early metabolicprogramming. *Bundesgesundheitsblatt, Gesundheitsforschung, Gesundheitsschutz, 53*(7), 666–673. doi:10.100700103-010-1079-y PMID:20652481

Letter of the Ministry of Health of Russia dated 24.03.2017 No. 28-1/10/2-1994 On the direction of guidelines "Recommended norms of medical nutrition (average daily sets of basic food products) for pregnant and lactating women in maternity hospitals (departments) and children of various age groups in children's hospitals (departments) of the Russian Federation".

Molloy, A. M., & Scott, J. (2001). Folates and prevention of disease. *Public Health Nutrition, 4*(2B), 601–609. doi:10.1079/PHN2001144 PMID:11683553

Muraro, A., Werfel, T., Hoffmann-Sommergruber, K., Roberts, G., Beyer, K., Bindslev-Jensen, C., Cardona, V., Dubois, A., duToit, G., Eigenmann, P., Fernandez Rivas, M., Halken, S., Hickstein, L., Høst, A., Knol, E., Lack, G., Marchisotto, M. J., Niggemann, B., Nwaru, B. I., ... Akdis, C. A. (2014, August). EAACI Food Allergy and Anaphylaxis Guidelines Group. EAACI food allergy and anaphylaxis guidelines: Diagnosis and management of food allergy. *Allergy*, *69*(8), 1008–1025. doi:10.1111/all.12429 PMID:24909706

Nazamova-Baranova, L. S. (2011). Allergy in children: bedside approach. Academic Press.

Robertson, A., Tirado, C., Lobstein, T., Jermini, M., Knai, C., Jensen, J.H., Ferro-Luzzi, A., & James, W.P. (2004). Food and health in Europe: a new basis for action. *WHO Reg Publ Eur Ser.*, (96).

Sicherer, S. (2001). Clinical implication of cross-reactive food allergens. *The Journal of Allergy and Clinical Immunology*, *108*(6), 881–890. doi:10.1067/mai.2001.118515 PMID:11742262

Thayer, Z.M., Rutherford, J., & Kuzawa, C.W. (2020). The Maternal Nutritional Buffering Model: an evolutionary framework for pregnancynutritionalintervention. *Evol Med Public Health,* (1), 14–27. doi:10.1093/emph/eoz037

The national program for optimizing the feeding of infants in the Russian Federation. (2010). Union of Pediatricians of Russia.

World Health Organization. (2016). *Mother's healthy diet: A better start to life*. WHO.

ENDNOTE

1 Currently JgE-mediated reactions are 3 times less common than non-JgE-mediated and mixed one.

Chapter 4
Fast Food:
Pros and Cons – Volumes of Consumption of Fast Food in the World and the Russian Federation: Fast Food Structure

ABSTRACT

The chapter is devoted to the analysis of modern fast-food products and an analytical overview of global fast-food enterprises. The authors substantiated the need to classify fast-food products and proposed their own approach. The authors' classification is based on combining linking features and dividing products into groups, taking into account production methods and storage methods, the degree of preparation for consumption, methods, and place of consumption. This classification combined technological approaches and fast-food products produced in public catering. The chapter also describes the leading global food corporations: fast-food catering and companies offering products and dishes of instant preparation. Brief characteristics of manufactured products, production technology, and other features are given. The purpose of this chapter was to analyze the spread of the impact of globalization on the food culture and fast-food consumption in the world.

DOI: 10.4018/978-1-7998-8197-1.ch004

INTRODUCTION OF THE CHAPTER

Food production and modern food technologies, unfortunately, are not a public health problem. Food processing and extension of their shelf life contributed to the evolution, adaptation and increase in the number of humanity and a settled population (Ludwig DS, 2011, Wrangham R., 2013, Pollan M., 2013, Kutkina M.N., Eliseeva S.A., 2018).

The need to increase food production has led to the mechanization of food production processes. Already from the middle of the 19th century, the mass production of some food products, such as bread, cookies, cakes, pies, sauces and meat products is marked. This was accompanied by a steady shift in emphasis in the habitual (traditional) diets of the population.

In the 20th century, standardized and unified technologies led to an increase in the production of cheaper foods with a high content of fat and sugar. The consumption of such products provoked an increase in the growth of cardiovascular diseases, first in high-income countries and then throughout the world (Omran AR., 1971, Popkin BM, 1993).

Research in the field of food sciences in general, from the 1970s to the 1980s, was aimed at creating and expanding the range of products with a long shelf life without special conditions produced from cheap ingredients and food additives.The global development of production of such products, their worldwide distribution and intense consumption has led to a sharp increase in obesity and related noncommunicable diseases (diabetes), first in high- and middle-income countries (Popkin BM, 2002), and then in countries with low income (Popkin BM, 2013).Food supplies are now becoming part of the global food system, in which ready-to-eat processed products are increasingly dominant (FAO, 2004, Stuckler D.et al, 2012, Monteiro CA.et al, 2012, McDonald's Corporation, 2019).

The traditional perception of the term "fast food" is restaurants, cafes and other fast food enterprises, however, the concept of fast food in modern times should be considered much more broadly.

MAIN FOCUS OF THE CHAPTER

4.1 Classification

There are different directions in the classification of fast food products, but no one classification is complete. This is primarily due to the huge variety of assortment and classification approaches. The main unifying feature of these classifications is the speed of consumption and (or) food technology. Fast food can be divided into 3 categories:

- category 1 includes street fast food, mobile stalls aimed primarily at consumption on the go (walking, sightseeing tour, working snack) and often owned by private entrepreneurs. Potential hazards having hygienic and technological components can be associated with this type of food;
- category 2 includes enterprises offering quick consumption and service. Such services are offered by global brands recognized in the world. The features of their activities are unified technological solutions that are economically feasible for the manufacturer;
- category 3 includes enterprises that produce products that can be quickly or instantly cooked, that is, that do not require special expenses of time and resources. These are semi-finished products of varying degrees of readiness, ready-made dishes with a long shelf life, concentrated products. Production technologies are fully industrial, and products, due to processing methods and long shelf life, often lose their native properties and essential substances.

Based on this formulation of the question, a much larger number of products fall under the classification of fast food products.

In industrial technologies, when producing a fast-food product similar to a traditional product, but with a long shelf life, several problems are solved in the process of product design, which ensure stability during storage. This often happens due to the replacement of natural ingredients with nutritional supplements.For example, in the production of frozen semi-finished products from vegetables such as cutlets, a chicken egg (which is not only a binding component, but also increases the biological value of the product) is replaced with a polysaccharide (methyl cellulose or various gums, etc.), which does not carry any food load, but it can act as a ballast substance that binds not only contaminants, but also the necessary substances.An important task of fast food technology is to predict the consequences of regular consumption.

That is why the authors offer their own approach to the classification of modern foods intended for fast consumption.

Table 1. Classification of global fast food products

The basis for grouping	Group	Group characteristics	Group signs	Product examples	Comments
According to the purpose	Instant	For instant consumption	It only needs to be unpacked before consumption	Confectionery, appetizers (chips, nuts, sandwiches, burritos, hamburgers, etc.)	It can be consumed anywhere; cutlery, hand washing, etc. are not required.
			Minimal effort for food regeneration	All products in which it is necessary to either add water or warm up	
	Requiring food regeneration	(At home, or in catering)	Requiring cooking and (or) recovering after freezing	Semi-finished products of a high degree of preparation (stuffed pancakes, etc.)	Analogies of traditional dishes with adjusted composition
			Full heat treatment cycle	Chilled and frozen convenience foods	Cooking time depends on the size of the product and its type, as close as possible to traditional
By product type	Appetizer	Snacks	Do not require special storage and preparation conditions	chips; cheese snacks; meat snacks; sticks; cookies; muesli; crispy flakes; nuts; dried fruits; chocolate bars etc.	Subdivided into sugarless and sweet snacks. Healthy snacks, such as yoghurts, cottage cheese, dairy and sour-milk drinks in small packages, are positioned as snacks to satisfy a slight hunger.
		Cold dishes and appetizers	Require special storage equipment, and special storage conditions	salads; burgers, sandwiches with minced meat, chicken, fish	Often, raw materials are produced (prepared) using standardized technologies
	Entree	Concentrated sidedish	Do not require special storage equipment, and special storage conditions	Noodles, potato puree, cereals	
		Second courses of high degree of preparation	Require special storage equipment, and special storage conditions	Pilaf, pizza, nuggets, chops, etc.	

continued on following page

Table 1. Continued

The basis for grouping	Group	Group characteristics	Group signs	Product examples	Comments
	Soup	Concentrated	Do not require special storage equipment, and special storage conditions	Broths, concentrate soups requiring cooking	
		Of high degree of preparation	Require special storage equipment, and special storage conditions	Cook & Chill technology	
	Desserts	Concentrated	Do not require special storage equipment, and special storage conditions	Jelly, creams, mousses, kissels	
		Of high degree of preparation	Require special storage equipment, and special storage conditions	Jelly, creams, mousses, kissels	
According to the features of technological processing	By storage method	Chilled	Stored at t = -2…4°C. Short shelf life from 1 to 21 days, depending on technology	Semi-finished products, ready meals and snacks	
		Frozen	Stored at t = -18°C Shelf life from 4 months to 1 year	Vegetable mixtures, eat, fish, poultry, vegetables semi-finished products of high degree of preparation	semi-finished products of a high degree of preparation, suggest heating the product to the desired temperature, combined with defrosting
		Canned/ambient/ shelf stable	Stored without special conditions. Shelf life up to 2 years	All types of canned food that have undergone sterilization	
		Dried/dehydrated	Shelf life from 6 months to 1 year	Instant concentrates and requiring additional heat, as well as dry snacks	
	According to the main (prevailing) technological methods / processing features	Fried in a lot of fat. Fryingtechnology	The product has a crisp and absorbs a large amount of fat; shelf life depends on the purpose of the product	Chips, Frenchfries, donuts	
		Dehydrated. Dehydration technology	Brittle, with a long shelf life		
		Combined. Combines several collaborative or sequential techniques	Ultra-processed, with refined ingredients	Semi-finished products of high degree of preparation, fried before freezing	Often have formed taste and aroma.

4.2 Leading Global Food Corporations

4.2.1 Fast Food Catering

By the 21st century, several of the largest fast food companies, holding steady leading positions have been established. Corporations operate under franchising conditions and develop their networks.

Table 2 presents comparative data on the distribution of business concepts (franchises) of the three main global fast food market players and the number of enterprises of these corporations in different countries of the world.

Table 2. Main global fast food market players and the number of enterprises of these corporations in different countries of the world

Country	Number of enterprises		
	McDonald's	Subway	KFC
USA	13 914	23882	4491
Japan	2975	225	1200
PR China	2700	533	4563
Germany	1480	695	
Canada	1450	3127	200
France	1419	412	
Great Britain	1274	2429	784
Australia	920	1305	600
Brazil	812	1868	
Russia	649	529	918
Philippines	640		
Italia	587		
Spain	509		
RepublicofKorea	447	375	
Poland	427		
Taiwan	413		
Mexico	402	898	
SaudiArabia	304		
India	290	663	299
Malaysia	282	233	579
RepublicofSouthAfrica	275		
Turkey	253		

continued on following page

Table 2. Continued

Country	Number of enterprises		
	McDonald's	Subway	KFC
Netherlands	249	214	
Thailand	240		
HongKong	237		
Argentina	222		
Austria	195		
Sweden	191		
Israel	185		
UAE	172		
Indonesia	170		466
Switzerland	167		
NewZealand	166	251	100
Portugal	148		
Singapore	136		
Venezuela	133		
Egypt	114		
CzechRepublic	99		
Guatemala	95		
Hungary	89		
Denmark	89		
Ireland	89		
Belgium	85		
Columbia	81	297	
Romania	80		
Ukraine	80		29
Kuwait	77		
Chile	77		
Norway	71		
Finland	65		
Pakistan	60		
Panama	57		
CostaRica	54		
Morocco	47		

*From Wikipedia: McDonald's, Subway, KFC

Obviously, the companies listed in Table 2 have different market impacts; McDonald's has a stable market share in all countries and provides more comprehensive reports on its activities. Subway has the largest number of points in the USA, almost twice as much as McDonald's. At the same time, distribution is not so stable throughout the world, although in Canada, Great Britain, Australia and Brazil, Subway has a significant advantage over McDonald's in terms of the number of enterprises.

The number of KFC in the USA is 3 times less than of McDonald's, and 5.3 times less than of Subway, while China has almost the same number of KFC as in the USA, which is 2 times more than of KFC. It should also be noted the high popularity of KFC in Indonesia, Malaysia, Japan and Russia. KFC is the most export oriented company.

There are several other well-known companies actively exploring the global fast food market, such as Burger King, Wendy's, Taco Bell, Chick-fil-A, Chipotle Mexican Grill, Sonic America's Drive-In, Dairy Queen and others.

Based on the data in Table 2, the food market has a very strong food culture.According to the authors, a clear understanding is needed of how regular consumption of fast food affects the health and healthy lifestyle of a modern person. It can be achieved by the analysis of the assortment and technological solutions used in this segment of fast food.

McDonald's is the most recognizable brand today. It has been around for 75 years.

The assortment is based on hamburgers ("Big Mac"), French fries, fried sweet cakes, and a traditional Cola drink. Moreover, the assortment of restaurants, the size and composition of servings can vary greatly in different countries, taking into account ethnic preferences.

The features of the Big Mac hamburger are that the bun is cut into three parts (the top part is sprinkled with sesame seeds on top). This stabilizes the hamburger. The filling is two cutlets weighing approximately 50 ($\pm$ 10) grams each, cheese, slices of pickled cucumber, onions, lettuce and Big Mac sauce.

Table 3 presents the nutritional and energy value of standard Big Macs in different countries. Obviously, for fat content, one Big Mac covers from 37 to 57% of daily amounts adopted in different countries. For protein, the need of the human body can also be met by 32 - 45%, depending on daily amounts adopted in different countries. Significant differences are observed in the amount of salt, probably, depending on preferences, the difference is more than 2 times.

Table 3. Comparative nutritional value of standard Big Macs in different countries (% of the guideline daily amounts)

Country	Energy value. kcal	Carbonhydrates, g	Protein, g	Fats, g	Fiber, g	Salt,mg	Weight, g
Argentina	495	42 (14%)	24 (32%)	26 (47%)	2.5 (10%)	1083 (45%)	202
Australia	480	36.2	25.3	24.9		800	201
Brazil	504	41 (14%)	25 (33%)	27 (49%)	3.5 (14%)	1023 (43%)	
Canada	540	44 (15%)	24	29 (45%)	3 (12%)	1020 (43%)	209
Croatia	495	40 (15%)	27 (36%)	25 (37%)	3 (12%)	2300 (46%)	219
CzechRepublic	495 (25%)	40 (15%)	27 (36%)	25 (37%)	3 (12%)	2300 (46%)	219
Denmark	497	43	27.1	24.1			219
Estonia	495	40	27	25		2300	219
Finland	495	40	27	25	3	2300	219
France	492	38.9	26.2	25.8	4.2	900	
German	495	40	27	25	3	2300	221
Greece	490	42.5	27	24	5	2000	
HongKong	510	39	25	28		870	
Венгрия	495 (25%)	40 (15%)	27 (36%)	25 (37%)	3 (12%)	2300 (46%)	
Italy	505	43	27	25	4		
Japan	545	42.7	26.1	25.8	2	864	216
Malaysia	484	46	26	23		730	209
Mexico	600	50	25	33	4	1050	219
Netherlands	495	40	27	25	3	2300	
NewZealand	464	33.9	24	25.4		1020	195
Norway	495 (25%)	40 (15%)	27 (36%)	25 (37%)	3 (12%)	2300 (46%)	
Poland	495	40	27	25	3	2300	
Russia	495	41	27	25	3	2300	208
RepublicofSouthAfrica	559	49.3	27.7	24.89	4.6	801.98	
Ukraine	503	42	26	25		2200	
RepublicofKorea	535	46 (14%)	27 (45%)	29 (57%)		750 (22%)	219
Sweden	495	40	27	25	3	2300	
Switzerland	495	40	27	25	3	2300	
Great Britain	493	44	26.7	22.9	5.9	2000	
USA	540	45 (15%)	25 (45%)	29 (45%)	3 (12%)	1040 (43%)	214

**McDonald's,fromWikipedia*

Based on the analysis of Table 3, a reasonable question arises - if one hamburger for fats and proteins makes up 1/3 of the daily norm, then is there a skew in the diet per day?

Technology: deep frying with a small amount of fat (for hamburgers).

Technologies are standardized and unified, this allows one to control and maintain a recognizable franchise. That is, a fully industrial production of cutlets, buns for hamburgers, other components of hamburgers is launched. It is these technologies that ensures quick cooking and realization of products.

In fact, at the enterprise that serves the visitor, only the last processing of the product takes place before implementation.

It should be noted that McDonald's has been criticized as an example of globalization for a long time. It is often shown as a negative example of globalization, since flavorants, stabilizers, preservatives, dyes, etc. are used in food production (John F, 2007, Horovitz, 2014).

Subway is a network of fast-food restaurants, also operating on the principle of franchising. Subway positions itself as a healthy alternative to standard fast-food restaurant chains, which is reflected in the slogan of the company "Eat Fresh!"

A distinctive concept is that the client chooses the type of bread for his sandwich, the type of sandwich (the main ingredient) itself and additional ingredients like cheese, vegetables and sauce.

Assortment: the main sold products are sandwiches, salads.

Subway offers sandwiches in three sizes:

- 30-cm sandwich (Eng. Footlong size);
- 15-cm sandwich (Eng. 6 inch; usually prices are indicated for these sandwiches);
- 10-cm mini-subs.

Despite the fact that there are no deep-fried foods in Subway, all sandwich ingredients and bread are industrial products.

Subway has been repeatedly criticized for the lack of strict quality control of the used ingredients and the service quality. They note that different restaurants have different service level. In one restaurant, service can take 3-7 minutes, in another - 20 minutes or more. The quality of sandwiches varies greatly, as some entrepreneurs use good quality products, while others reduce costs by using cheaper products, for example, soy analogues or imitation of crab sticks in a seafood sandwich instead of ham. Most restaurants have significantly different sandwich menus.

Burger King Corporation is an American company, the owner of a global chain of fast food restaurants Burger King, specializing in hamburgers (mainly Whoppers).

Standard Whopper is a hamburger, consisting of a flame grilled 4 oz (110 g) beef patty, sesame seed bun, mayonnaise, lettuce, tomato, pickles, ketchup, and sliced onion. Optional ingredients such as American cheese, bacon, mustard, guacamole or jalapeño peppers may be added upon request. Barbecue, salsa and guacamole sauces are also available in some countries. In addition, Burger King adds to Whopper any of the seasonings sold in this restaurant upon request.

Whopper is a higher-calorie burger than the Big Mac: 670 kcal versus 540 kcal, respectively. At the same time, Whopper is bigger than Big Mac: 290 g versus 214 g. Thus, Whopper contains less kilocalories per gram: 231 kcal / 100 grams versus 252 kcal / 100 grams, respectively. This difference is not constant, as in Big Mac cheese is used by default, and in Whopper - upon request.

As of March 2012, about 12.5 thousand Burger King restaurants were operating in 76 countries of the world, of which over 90% were franchised.

In Australia, a Burger King franchise operates under its own name, Hungry Jack's, named after its owner, Jack Covin.

Wendy's - the chain has 6207 franchised restaurants around the world (as of January 2017), 77% of which are located in North America. It is the main competitor of the Burger King, Subway and McDonald's.

The assortment of Wendy's restaurants includes 10 types of burgers made of beef, 9 types of chicken fillet burgers, several types of French fries, baked potatoes, 5 types of salads, Chile specialty stew and about 30 desserts and drinks. In addition, the range is updated every season.

Taco Bell is an international chain of Tex-Mex adapted fast food restaurants. Currently, the Taco Bell chain has about 6.4 thousand restaurants. Most of the institutions (5.5 thousand) is franchised. The company owns 0.9 thousand restaurants. The main market for the chain is locatedin the United States, where 6.2 thousand of its restaurants operate. Therefore, Taco Bell remains predominantly an American fast food chain. However, the company plans to have an active international expansion. As a result, by 2023, the number of restaurants should increase to 8 thousand. At the same time, Taco Bell twice unsuccessfully tried to enter the Mexican market.

Assortment: adapted Tex-Mex cuisine: burritos, tacos, nachos, Mexican pizza, etc.

It should be noted that Mexican adapted cuisine is popular in the United States. In North America, there are a number of other fast food chains with the word Taco in the name: Taco Bueno, Taco Cabana, Taco del Mar, Taco John's, Taco Mayo, Taco Palenque, Taco Tico, Taco Time, Del Taco, etc.

Chick-fil-A is an American fast food restaurant chain specializing in chicken sandwiches. The chain of enterprises rather small in comparison with others totals more than 1850 restaurants.

In advertising, the company claims that no trans fats are used in cooking, as well as chickens raising using antibiotics.

Kentucky Fried Chicken (KFC) is an international restaurant chain specializing in chicken dishes.

The basis of the KFC menu is pressurized, seasoned fried chicken slices - Original Recipe.

Then the product is frying under pressure at a temperature of 185 °C; the maximum frying time is 7 minutes (Balakrishnan, M. Stephens., 2013) After this, the product cools for 5 minutes and is placed in a furnace operating in the heating mode. If a product is not consumed within an hour and a half, then, according to company policy, it is discarded. In different countries, sunflower, soybean, rapeseed or palm oil can be used for frying (Steyn L.,2013).

Other key products of the company are chicken burgers, including Zinger and the Tower series, wraps (Twister and Boxmaster series) and dishes eating with fingers: crispy chicken "strips" and spicy wings. Popcorn Chicken is one of the most common KFC products; the dish consists of small pieces of fried chicken. Chicken nuggets are sold in some regions, and in some markets, such as Australian, the product is known as Kentucky Nuggets (David E., 2011).

KFC adapts the menu for different countries, so the total range of cafes includes more than 300 items.

McCormick & Company is the largest supplier of KFC sauces, condiments and pickles. At the same time, the companies collaborate in the development of new dishes.

Due to the long history of relations with PepsiCo, in most countries the cafe delivers drinks from this particular manufacturer. Nevertheless, the Coca-Cola Company supplies product in the Republic of South Africa, Turkey, Romania, Greece, the Philippines and Barbados; Cott Corporation RC Cola is available to visitors in Aruba. In Peru, KFC restaurants sell Inca Kola, popular in the country.

In 2009, the company launched the Krusher / Krushem line of frozen drinks containing pieces of Kit Kat, Oreo snacks or strawberry shortcake snacks. Frozen drinks are available at over 2000 cafes (Jargon J.,2012). Custard egg tart is a popular KFC dessert all over the world, and in Peru, cafe visitors can try Sunday and Tres leches (Three Milk) cakes. In 2012, a morning menu "KFC am" appeared in a number of countries, which, in addition to the standard chicken, included pancakes, waffles and porridge (Bowman J., 2014).

Chipotle Mexican Grill, Inc. is a chain of restaurants in the United States, Britain, Canada, Germany and France, specializing in burritos and tacos cooking. The name of the company comes from the Mexican seasoning chipotle, which is a smoked red jalapeno pepper.

According to The Motley Fool, in October 2014, Chipotle had 17 cafes outside the United States, most of them were located in Canada and the United Kingdom (Dean T., 2013).

The Chipotle assortment consists of four items: buritos, dishes on plates, tacos and salads. Dishes are formed at the client's choice: chicken, pork carnitas (available only in some restaurants), sofritas based on tested tofu, barbacoa, steaks or guacamole (vegetarian content). Additional toppings include rice, beans, four kinds of salsa sauce, sour cream, cheese or lettuce.

Chipotle offers vegetarian menu: rice, black beans, fajita vegetables (onions and sweet peppers), salsa, guacamole and cheese. All items except meat, cheese, sour cream and honey sauce belong to the vegan diet.

The company declares that it basically cookes only from natural and ethically flawless products. The vast majority of food is cooked in every restaurant, with a few exceptions in the form of beans and carnitas, which are cooked in a central kitchen in Chicago, Illinois. None of the company's restaurants have refrigerators, microwaves, or can openers (Carter, R., 2008).

At the end of 2013, the company developed a new cooking strategy for spotted beans, which eliminated bacon and made it possible to prepare food for vegetarians and vegans (Tepper R.,2013). The cheese was processed on the basis of rennet to be suitable for vegetarians. In April 2010, Chipotle, at six locations in the United States, began testing a vegan version of Garden Blend based on industrial-made adobo pickled meat substitutes. Flour tortillas used for burritos and soft tacos and chips were the only gluten containing items.

In 2001, Chipotle recalled a company mission called Food With Integrity, highlighting Chipotle's efforts to increase the proportion of naturally-grown meat, organic food, and hormone-free milk. Chipotle uses only chicken legs; the breast is sold to Panera Bread (Chipotle Mexican Grill, From Wikipedia).

Criticism and reviews: In 2003, Center for Science in the Public Interest reported that Chipotle burritos contain over a thousand calories, corresponding to two servings of food. The American channel MSNBC Health.com has listed burritos in the "20 Worst Foods in America" List because of its high calorie and sodium content. While burritos with carnitas, rice, vegetables, cheese, guacamole and salsa are comparable to Big Mac, burritos contain more fat, cholesterol, carbohydrates and sodium than Big Mac, but at the same time it contains more proteins and fibers. The restaurant also received Health.com feedback, which included the restaurant in the list of "fastest-food restaurants with the healthiest foods" (Chipotle Mexican Grill, From Wikipedia).

In 2008, 400 people in Kent, Ohio caught Norovirus and became ill after visiting Chipotle Restaurant.

Sonic America's Drive-In - snack bars at gas stations, about 3,500 points in number. Beloved by American truckers and drivers, a drive-in chain with waitresses on roller skates, hamburgers and French fries, onion rings and hot dogs.

Dairy Queen - about 5 thousand points for the sale of soft ice cream. Popular among teens.

Assortment: burgers, ice cream.

It should be noted that Italian cuisine, which has long been adapted to world food, is also very popular.

Pizza Hut - a chain of restaurants, divided into two types of restaurants: Pizza Hut for a family dinner and fast food Pizza Hut Express. Family pizzerias differ from fast foods in a more diverse menu, more places and more staff (as, for example, in the Golden Corral restaurant chain). It is the largest pizza chain in the world, specializing in the cooking of the American version of pizza, as well as side dishes such as pasta, chicken wings, bread rolls, garlic bread.

Domino's Pizza is an American catering company. It manages the world's largest chain of pizzerias (Pizza Hut is inferior in the number of restaurants). The chain, represented in 85 countries and including 15,900 restaurants, sells more than 3 million pizzas a day. The company owns 390 restaurants in the United States, the rest operate on franchising.

The main market for the company is located in the United States. The pizza segment is divided into service in pizzerias, carryout pizza and delivery. The main competitors in the USA are Pizza Hut, Papa John's and Little Caesars Pizza, in the international arena - Pizza Hut and Papa John's.

The main units (data for 2018) are presented below.

Restaurants in the USA include 390 own restaurants in the USA and 5486 franchising employees. The largest number of Domino's pizzerias are located in India (1195), Great Britain (1100), Mexico (760), Australia (693), Japan (550), Turkey (535), Canada (487), Republic of Korea (447), France (387) and Germany (283). In foreign markets, Domino's Pizza works through large master franchisees, which either manage pizza chains themselves or are engaged in sub-franchising. (From Wikipedia).

Supply: provision of dough, vegetables and other convenience foods, as well as of equipment for Domino's pizzerias in the USA and Canada. This division accounts for 57% of the company's revenue; 20 production centers are located in the USA and 5 in Canada; in 2018 they produced 260 thousand tons of dough.

Many corporations are already combining several brands of fast food. For example, an American catering corporation **Yum! Brands, Inc.** It is included in the list of 500 largest US companies, according to Fortune magazine (in 2019 at 494th place). Among others, Yum! owns 3 world brands - KFC, Pizza Hut and Taco Bell. In 2016, activities in China were spun off into an independent company, Yum China Holdings, headquartered in Louisville, Kentucky.

To understand the scale of globalization, the statistics of the main units (data for the end of 2018):

KFC - 22,621 fast food restaurants in 136 countries (18% of restaurants are located in the USA); the total turnover of this chain is $ 26.2 billion, the share of franchising is 99%, so only royalties are due from companies with this amount.

Pizza Hut - 18,431 pizzerias in 111 countries (41% in the USA); total turnover is $ 12.2 billion; the share of franchising is also 99%.

Taco Bell - 7072 restaurants with Mexican cuisine; located in 27 countries, but 93% of restaurants are located in the United States; the total turnover is $ 10.8 billion, the share of franchising is 93%.

Summing up the characteristics of fast food restaurants and assortment, it should be noted that the main players in the fast food market are huge corporations with industrial enterprises and unified technologies. This has both positive and negative features.

The positive ones include control over the technology for the production of semi-finished products, ensuring the safety of HACCP systems in industrial production, and monitoring compliance with standards in franchised enterprises.

The negative features are high-calorie food, lack of balance in essential substances, aggressive food processing technologies, and as a result, the effect of regular consumption on the human body.

4.2.2 Companies Offering Products and Dishes Of Fast And Instant Cooking

The trend towards fast food consumption is manifested not only in the development of the fast food chain. There is another segment, designed for consumption at work or at home and involving minimal effort for cooking. These include soups, main courses such as noodles and instant potatoes, and some types of sweet foods and pastries. A special place is occupied by instant products, designed to minimize labor and energy costs.

Currently, according to the World Association of Instant Noodles (WINA), the most popular concentrate is instant noodles (Table 4).

Table 4. Global demand for instant noodles,million servings

Region,country	2013	2014	2015	2016	2017
China /Hong Kong	46.220	44.400	40.430	38.52	38.97
Indonesia	14.9	13.43	13.2	13.01	12.62
Japan	5.52	5.5	5.54	5.66	5.66
India	4.98	5.34	3.26	4.27	5.42
Vietnam	5.2	5	4.8	4.92	5.06
USA	4.35	4.28	4.08	4.1	4.13
Philippines	3.15	3.32	3.48	3.41	3.75
Republic of Korea	3.63	3.59	3.65	3.83	3.74
Thailand	3.02	3.07	3.07	3.36	3.39
Brazil	2.37	2.37	2.37	2.35	2.23
Russia	2.12	1.94	1.84	1.57	1.78
Nigeria	1.43	1.52	1.54	1.65	1.76
Nepal	1.02	1.11	1.19	1.34	1.48
Malaysia	1.35	1.34	1.37	1.39	1.31
Mexico	0.92	0.900	0.850	0.890	0.960
Taiwan	0.750	0.710	0.680	0.770	0.820
Myanmar	0.340	0.410	0.460	0.570	0.590
SaudiArabia	0.660	0.490	0.510	0.510	0.530
Australia	0.350	0.360	0.370	0.380	0.390
Great Britain	0.370	0.360	0.360	0.360	0.360
Bangladesh	0.220	0.250	0.270	0.290	0.350
Cambodia	0.240	0.250	0.270	0.280	0.330
Ukraine	0.580	0.580	0.410	0.280	0.330

continued on following page

Table 4. Continued

Region,country	2013	2014	2015	2016	2017
Poland	0.260	0.300	0.310	0.310	0.310
Guatemala	0.080	0.210	0.190	0.270	0.270
SouthAfrica	0.170	0.190	0.190	0.200	0.240
Egypt	0.170	0.190	0.200	0.210	0.220
Kazakhstan	0.140	0.140	0.140	0.240	0.210
German	0.180	0.180	0.190	0.200	0.200
Canada	0.210	0.190	0.190	0.190	0.190
Pakistan	0.130	0.150	0.160	0.170	0.190
Peru	0.080	0.120	0.130	0.130	0.140
Singapore	0.130	0.130	0.130	0.130	0.130
Uzbekistan	0.110	0.110	0.110	0.100	0.120
New Zealand	0.070	0.080	0.080	0.080	0.080
CzechRepublic	0.090	0.060	0.090	0.090	0.080
Ethiopia	n/a[1]	0.040	0.050	0.040	0.080
Spain	n/a	0.040	0.050	0.070	0.070
France	0.060	0.060	0.060	0.060	0.060
India	n/a	0.020	0.030	0.040	0.050
Iran	n/a	0.040	0.040	0.040	0.050
Kenya	n/a	0.040	0.040	0.050	0.050
Sweden	0.030	0.030	0.030	0.040	0.040
Hungary	0.020	0.020	0.020	0.030	0.030
Netherlands	0.020	0.020	0.020	0.020	0.030
Chile	0.030	0.040	0.040	0.030	0.020
Italy	n/a	0.010	0.020	0.020	0.020
Belgium	0.010	0.020	0.020	0.020	0.020
Columbia	0	0.010	0.010	0.010	0.010
CostaRica	0.010	0.010	0.010	0.010	0.010
Denmark	0.020	0.020	0.010	0.010	0.010
Finland	0.020	0.010	0.010	0.010	0.010
Switzerland	n/a	0.010	0.010	0.010	0.010
Argentina	n/a	0.010	0.010	0.010	0.010
The rest	0.250	0.940	0.940	0.930	1.15
In total	105.99	103.96	97.51	97.47	100.1

* the World Association of Instant Noodles (WINA)

According to ACNielsen, in 2008, sales grew in value terms in all segments of the instant food market. Fast food products have become a familiar part of the modern consumer culture. According to experts, this market (especially the noodles / instant noodles segment) is at the peak of its saturation: it is becoming increasingly difficult for manufacturers to find promising niches and track current demand trends (LiveJournal 2013).

According to the study by Business Rating agency, the market of instant products is a fairly dense market in which there are no more “empty places” and there is a fierce competition between manufacturers.

This fact is also confirmed in a report published in those years by MAREVEN FOOD CENTRAL (TM “Rolton”). According to the report, the market, showing stable growth before, stopped at around 1.1 billion. Many analysts agreed with these conclusions, arguing that the FSN market is close to saturation (Shin H.J.et al,2014).

According to the World Instant Noodle Association, more than 46.2 billion packages were sold in 2013 in China and Hong Kong, but by 2017, these sales fell to 38.97 billion packages (by 17%).

One of the reasons of this fact is that some Chinese consumers are raising their expectations in the dining room. Zhao Ping, a member of the China Academy’s International Trade Promotion Board said that the decline in sales of instant noodles indicated a change in consumption patterns in China. Consumers were more interested in quality of life than just filling their stomach.

Russia ranks 11th in the consumption of instant noodles in the world.

It should be noted that the popularity of this product is high in countries with a high level of per capita income.

The dynamics of consumption over the years does not show a significant decline in demand for this type of product.

Unfortunately, there are no reliable statistics on world consumption of all concentrates. At the same time, one can indirectly judge the popularity by statistics in Russia (Table 5).

Based on the data in Table 5, we can note a significant increase in the consumption of breakfast cereals (products from corn and other types of grain). The transformation of the consumption culture of these products by 2-2.5 times compared with 1990 is obvious. An increase in the share of imports by 2013 (36.5% compared to 4.0% in 1990) also indicates a borrowing of global consumption trends. At the same time, a decrease in imports over the past 7 years and an increase in domestic production evidence the popularity of this group of concentrates in Russia, confirming the influence of global processes in the nutrition of the population.

Table 5. Dynamics of development of production, consumption and import of food-concentrate products in Russia

Product	Development period				Forecast
	1990	2013	2014	2015	2020
Food concentrates (Including instant ones)					
Consumption, t	1670	1120	1050	1090	1200
Production, t	1415	840	820	840	960
Importratio,%	15.0	25.0	22.0	23.0	20.0
Dry baby food					
Consumption, t	200	215	215	210	220
Production, t	169	97	97	105	130
Importratio,%	15.5	55.0	55.0	50.0	40.0
Breakfast cereals (products from corn and other types of grain)					
Consumption, t	451	1040	1090	1090	1300
Production, t	433	660	700	710	970
Importratio,%	4.0	36.5	36.0	35.0	25.0
Coffeeand coffeeproducts					
Consumption, t	307	1380	1390	1380	1500
Production, t	292	740	750	760	900
Importratio,%	5.0	46.0	46.0	45.0	25.0
Naturalpackaged tea					
Consumption, t	605	1560	1600	1660	1800
Production, t	575	810	830	830	990
Importratio,%	5.0	48.0	48.0	50.0	45.0

**Federal State Statistics Service 2016 Consumption of basic food products by the population of the Russian Federation*

The production and consumption of food concentrates (soups, main dishes, and sweet dishes), was well developed until 1990. A decrease in domestic production and an increase in the share of imports are directly related not only to the economic components of that period, but also to the continuity of new trends in consumption, especially to an increase in demand for instant concentrates.

The consumption of fast food products in various regions of Russia is presented in (Table 6).

The main share of instant food consumers are people aged 25-54. At the same time, the consumption of these products is typical for younger age groups (10-19 years).

Table 6. The consumption of fast food products in various regions of Russia

Region	The consumption of fast food products, % of the number of consumers			
	Soup	Vermicelli	Cereal	Potato puree
Moscow	52.6	51.9	6.7	4.8
Saint-Petersburg	35.3	71	2.3	3.7
EasternSiberia	27.9	73.5	4.2	2.2
Central Region	40.4	66.4	5.2	2.1
Ural	19.9	87.8	3.5	1.7
North / Northwest	32.2	74.6	3.9	1.5
FarEast	34.4	70	6.1	1.5
NorthCaucasus	31.8	71.5	3.5	1.3
Povolzhye	31.5	81.7	2.1	17
WesternSiberia	22.4	75.1	3	0.6
CentralBlackEarthRegion	40	64.6	1.3	0.4

**Federal State Statistics Service 2016 Consumption of basic food products by the population of the Russian Federation*

The consumption of soups and second lunch dishes of quick and instant cooking is typical for people with different levels of income. However, the socio-demographic characteristics of consumers of certain types of instant products are somewhat different. For example, cereal consumption is largely characteristic of high-income groups. It should be noted that instant products are consumed by both men and women. However, cereals are a predominantly "female" product: 2/3 of their consumers in Russia and in Moscow are women.

This type of product is considered as poor-quality and junk food, that is one of the main reasons for the decline in market growth. It led to a transformation in the market, a rapid re-positioning of the products. They have evolved from cheap and affordable to convenient and fast.

Consumer preferences in instant foods have changed significantly. Earlier, price was one of the main motives for buying the product, but now preferences have shifted towards saving time, not money. The acute relevance of the low price provided by local producers characterizes only regional demand. Consumers of megacities are more and more loyal to high-quality natural products with such additional characteristics as ease of cooking. It becomes the prospect of market development for the creation of instant products with high consumer value.

Currently, the most demanded products are these in high-quality packaging and with a complex composition of ingredients. The target audience is willing to pay for environmentally friendly packaging and a wider ingredient composition. With this focus on the eco product, one should ask whether the product itself meets the safety requirements with regular consumption.This question remains open also because the producer, controlling the quality and safety according to the criteria established in regulatory documents, does not focus on the transformation of food substances in the technological process (inside the product) and during storage.

Therefore, the trends in providing quality indicators for instant products are biased towards the improvement of packaging materials (briquetted products are becoming less popular than those packaged in glasses and bowls).

After analyzing the frequency of purchase of instant products, Business Rating Agency identified two popular answers indicated by 38% of respondents: “1-2 times a week” and “1 time per month”:

The main reasons for the consumption of this product, the respondents indicated the speed of cooking, as well as the lack of the ability to cook better food (59 and 43% of respondents, respectively).

Nestle Switzerland hold a strong position in the world in the production of food concentrates, it has several major brands such as Maggi (soup concentrates, seasonings and spices), KitKat, Nescafe and others.

In 2018, the main market for products was in the USA – $ 27.62 billion (almost a third of the company’s turnover), in China – $ 7 billion, France – $ 4.56 billion, Brazil – $ 3.68 billion, the UK – $ 2.93 billion, Mexico, Germany, the Philippines and Canada – from 2.0 to 2.8 billion, Italy, Japan, Russia, Spain, Australia, India, Switzerland – from 1.24 to 1, 82 billion, respectively.

The company owns 461 factories and industrial enterprises in 83 countries.

Currently, Nestle is a global corporation developing its enterprises in all regions (Table 7).

Table 7. Main product groups and sales in 2018

SalesRegions / divisions	Product turnover, % of the total turnover in the region			
	dairyproducts	semi-finished products and concentrates	instantdrinks	infantfood
Zone Americas	22.5	17.9		
Zone Asia, Oceania and Sub-Saharan Africa	24.1	12.2	28.5	22.9
Zone Europe, Middle East and Africa		20.7	27.2	

*Nestlé, from Wikipedia

In 2018, the turnover of soluble and liquid drinks estimates $ 21.6 billion, of infant and medical food – $16.2 billion, food concentrates and convenience foods – $12.1 billion, respectively.

Unilever company is the main competitor in the production and sale of food concentrates and other products, but according to the Global 500 List, the Nestle's income is 1.5 times more than that of Unilever;

Unilever is a Dutch-British company, it is one of the world leaders in the market of food products and household chemical goods (including perfumes).

German brand KNORR, owned by Unilever produces concentrates. The assortment includes soups, seasoning mixes and sauces.

- **Nissin Food Products** (Japan) is specialized on instant noodles production. Its offices and factories are located in the USA (since 1972), Brazil (since 1981), in Hong Kong, India, Hungary, Germany, Thailand, China, and Mexico. The company operates 29 enterprises in 11 countries. Products are sold in 16 countries.
- **Mareven Food Central** (Vietnam), the Rollton trademark. The company is represented in Russia, Ukraine, CIS countries, EU countries, Africa and Asia. Production of instant noodles, mashed potatoes, broths, side dishes and seasonings.
- **Dr. Oetker** (Germany). The activities of the company are nationally oriented. At the same time, production and representative offices are located in 40 countries, mainly in Western and Eastern Europe, as well as in North and South America.
- The main concentrate products manufactured by the company are desserts - fruit jellies, puddings, Panna Cotta and Tiramisu concentrates, the company also specializes in universal baking mixes, baking decor, baking ingredients and semi-finished products.
- **Cykoria SA** (Poland). The main recipients of products are Russia, Germany, France, Spain, Portugal, Italy, Great Britain, Switzerland, the USA, Ukraine, Croatia, the Czech Republic and Slovakia, but a significant part of the products is in demand on the Russian market. The main assortment of the company consists of seasonings and spices, concentrates of sweet dishes, and an assortment of jelly oriented to Russia and Ukraine.
- **Podravka** (Croatia). The sales markets are located in more than 40 countries: Europe, USA, China. The famous seasoning is VEGETA. The company produces concentrate of soups requiring cooking.

- **Gallina Blanca** (Spain). It occupies a leading position in the Spanish market for the production and sale of friable broths and bouillon cubes, instant soups, second course sauces, and seasonings. It operates in Europe, Central Asia, America and Africa.
- **Haas** (Austria) produces desserts - puddings, mousses, sweet sauces, kissels, creams, jellies, and baking mixtures.
- **BLA BAND** is the leader of the Scandinavian market of dry dishes for expeditions. The assortment includes cereals, goulash, Indian chicken broth, cereal with fruits, pasta Bolognese, etc. All products do not require boiling.

These are the largest global companies selling their products worldwide.

CONCLUSION OF THE CHAPTER

From the presented data, it is obvious that the consumption of fast food in the world is of a global nature. The network of fast food enterprises is constantly developing and modernizing, and the popularity of the products does not decrease. This fact should be taken into account in predicting the health of the population regularly consumption of fast food products.

In modern realities, taking into account the adequate consumption of essential nutrients, especially irreplaceable ones, is becoming an acute problem for mankind.

It should be noted that until now there was no uniform classification of fast-food products produced in public catering and industrial enterprises. The authors presented their own classification based on combined characteristics, which most fully reflects the entire range of fast food products.

The presented analysis of the prevalence of global catering enterprises in the world and the analysis of technological solutions used in these enterprises allows one to single out several unifying aspects for fast food products:

- the main ingredients that unite most fast food industries are fats and starch contained in various types of flour (wheat, corn, rice);
- the main technological methods for preparing products before selling or storing are frying or processing under pressure in combination with high temperatures.

- These ingredients and processing techniques used to prepare fast food products can produce substances that can adversely affect human health with regular consumption and dreary diet.

REFERENCES

Balakrishnan, M. S. (2013). East meets west: the world is round and time is cyclic. In M. S. Balakrishnan, I. Michael, & I. Moonesar (Eds.), *Actions and Insights - Middle East North Africa: East Meets West*. Emerald Group Publishing Limited.

Bell, D. E., & Shelman, M. L. (2011). KFC's Radical Approach to China. *Harvard Business Review*, *89*(11).

Bowman, J. (2014). *Why Chipotle Mexican Grill Stock Looks Like A Great Long Term Investment*. The Motley Fool.

Burger King Celebrates as the Whopper Turns 45. (2002). *Biz Journals*.

Carter, R. (2008). *Chipotle Mexican Grill to open Friday*. Record Herald.

Chicken. (2014). KFC Australia.

Dean, T. (2013). *Chipotle's Vegan Sofritas Now Served in All of CA*. VegNews.

Eldridge, D. (2014). *Moon Pittsburgh. Moon Handbooks*. Avalon Publishing.

FAO. (2004). Globalization of Food Systems in Developing Countries: Impact on Food Security and Nutrition. FAO Food and Nutrition.

Global Fast Food Market. (2021). https://ca.finance.yahoo.com/news/global-fast-food-market-2021-154100106.html?guccounter=1&guce_referrer=aHR0cHM6Ly95YW5kZXgucnUv&guce_referrer_sig=AQAAAMRvdD5BxftgmjNgajA-

Hesser, A. (2005). The Way We Eat. *The New York Times Magazine*.

Horovitz. (2014). Bruce McDonald's testing seasoned fries. *USA Today*.

Jargon, J. (2012, Feb. 21). Yum's CEO Serves Up New Taco, Growth Plans. *Wall Street Journal*.

KFC: Restaurants worldwide 2019. (2021). Statista.

Kutkina, M. N., & Eliseeva, S. A. (2018). Russian cuisine: From time immemorial to the present day. Academic Press.

Ludwig, D. S. (2011). Technology, diet, and the burden of chronic disease. *Journal of the American Medical Association*, *305*(13), 1352–1353. doi:10.1001/jama.2011.380 PMID:21467290

McDonald's Corporation - Company Profile, Information, Business Description, History, Background. (2019). NetIndustries, LLC.

Monteiro, C. A., & Cannon, G. (2012). The impact of transnational 'Big Food' companies on the South: A view from Brazil. *PLoS Medicine*, *9*(7), e1001252. doi:10.1371/journal.pmed.1001252 PMID:22802732

Moodie, R., Stuckler, D., Monteiro, C., Sheron, N., Neal, B., Thamarangsi, T., Lincoln, P., & Casswell, S. (2013). Profits and pandemics: Prevention of harmful effects of tobacco, alcohol, and ultra-processed food and drink industries. *Lancet*, *381*(9867), 670–679. doi:10.1016/S0140-6736(12)62089-3 PMID:23410611Omran, A. R. (1971). The epidemiologic transition: A theory of the epidemiology of population change. *The Milbank Quarterly*, *49*(4), 509–538. doi:10.2307/3349375 PMID:5155251

Pepitone, J (2011). *Subway beats McDonald's to become top restaurant chain*. Ulianne.

Pollan, M. (2013). *Cooked: A Natural History of Transformation*. Penguin Press.

Popkin, B. M. (1993). Nutritional patterns and transitions. *Population and Development Review*, *19*(1), 138–157. doi:10.2307/2938388

Popkin, B. M. (2002). An overview of the nutrition transition and its health implications: The Bellagio Meeting. *Public Health Nutrition*, *5*(1a), 93–103. doi:10.1079/PHN2001280 PMID:12027297

Popkin, B. M., & Slining, M. (2013). New dynamics in global obesity facing low- and middle-income countries. *Obesity Reviews*, *14*(Suppl. 2), 11–20. doi:10.1111/obr.12102 PMID:24102717

Shin, H. J., Cho, E., Lee, H. J., Fung, T. T., Rimm, E., Rosner, B., Manson, J. E., Wilin, K., & Hu, F. B. (2014). Instant noodle consumption and mode diets are associated with various cardiometabolic risk factors in Korea. *J Nutr. PubMed*, *144*, 1247–1255. PMID:24966409

State: Test results in on food samples taken from Kent Chipotle. (2008). WKYC.

Steyn, L. (2013). *KFC's secret recipe for growth. Mail & Guardian.*

Stuckler, D., McKee, M., Ebrahim, S., & Basu, S. (2012). Manufacturing epidemics: the role of global producers in increased consumption of unhealthy commodities including processed foods, alcohol, and tobacco. *PLoS Med, 9.*

Tepper, R. (2013). *Chipotle's Pinto Beans Will Soon Ditch Pork, Be Totally Vegetarian Nationwide*. Huffington Post.

Vegans and Carnivores Unite. (2015). Chipotle Mexican Grill.

Wrangham, R. (2013). The evolution of human nutrition. *Current Biology, 23*(9), 354–355. doi:10.1016/j.cub.2013.03.061 PMID:23660356

ENDNOTE

1 No Data Available.

Chapter 5
Study of Safety Indicators of the Most Popular Fast Food Products With a High Content of Fat Component

ABSTRACT

This chapter is devoted to the study of the processes of hydrolysis and oxidation of deep-frying fats and their influence on the safety indicators of the fat component of the most popular fast-food products. The studies were carried out by the authors of this work and under their guidance at the St. Petersburg State Technical University, St. Petersburg Polytechnic University of Peter the Great, and Saratov State Agrarian University named after N. I. Vavilov. Despite the trend towards consumption of healthy, low-fat foods, the popularity of deep-fried foods is significant throughout the world; the number of fast-food restaurants and the consumption of fast-food deep-frying products are increasing.

INTRODUCTION OF THE CHAPTER

The technology of deep-frying is becoming more widespread due to the high speed of heat transfer, simplicity and speed of cooking, attractive taste and the specific aroma of the finished product. The method of frying in deep fat

DOI: 10.4018/978-1-7998-8197-1.ch005

produces "fast food" products (French fries, chips, wings, burgers), as well as many dishes of national cuisine (belyash, cheburek, chak-chak, etc.).

The volume of production of fast food products in Russia is growing rapidly and, according to a number of marketing studies, exceeds 200 thousand tons per year. The annual growth of this market segment is 10-20%. The largest volumes of sales in fast food enterprises of the Russian Federation are accounted for by French fries - 30-35%, chicken wings - 15-20%, breaded chicken fillet - 10%, pies, belyash, cheburek - 10-15%, donuts - 15%, fish and fries seafood - 10%, fries vegetables - 5%, respectively [Ignatov V.I., 2010]. The consumption of special deep-fried fats and vegetable oils for the production of these products is at least 200 thousand tons per year (Nosova A.S., 2013.).

It is widely believed that deep-fried fast food products are unsafe for health. The problem is of particular importance also because the main group of consumers of these products is the young generation, for whom the negative effect of fat oxidation products on health is especially pronounced.

Clinical experiments on animals (white rats) revealed that the consumption of thermally oxidized frying fat has a negative effect on the digestive tract, changes the results of biochemical and clinical blood tests of experimental animals for the worse (Okunevich I.V., 2004; Simakova I.V., 2009; Makarova A.N., 2011).

For practical ensuring the safety of products, it is necessary to strengthen the input control of raw materials, improve the technology, control the parameters of the process, as well as to amend the current regulatory documentation for fast food products.

MAIN FOCUS OF THE CHAPTER

5.1 Methods for Monitoring the Safety Indicators of Fast-Fried Foods

On July 1, 2013, the Technical Regulations of the Customs Union TR CU 021/2011 "On Food Safety" were introduced. In accordance with the requirements of Article 10 of these regulations, food manufacturers must develop, implement and maintain procedures based on the principles of hazard assessment at critical control points of the technological process (Hazard Analysis and Critical Control Points - HACCP).

The analysis showed that the safety of fast food products is determined, first of all, by the degree of absorption of frying fat by the finished product and by the content of hydrolytic decomposition and thermal oxidation products formed in the technological process of raw materials' processing in this fat.

In this regard, the regulatory documents of the Federal Service for Surveillance on Consumer Rights Protection and Human Wellbeing provide for mandatory control of the products of thermal oxidation of deep-fat fats in the technological process (TR CU 021/2011, SanPiN 2.3.6.1079-01).

At the same time, the normative and technical documentation for fast food products does not contain specific safety indicators for these products that guarantee consumer safety.

Further development of the production of deep-fried products is possible only with the unconditional guarantee of the safety of these products.

In the production of fast food products, it is necessary to slow down the oxidation of frying fat as much as possible and, accordingly, limit the accumulation and the absorption of toxic oxidation products by the finished product.

At large enterprises, in the conditions of in-line production, for example, crackers and chips, it is possible to use complex technical means, including the use of inert gas, ensuring a minimum contact time of the raw material with hot deep frying oil and atmospheric oxygen.

In the conditions of numerous small enterprises and restaurants using the traditional technology of deep-frying, a number of methods were proposed for this purpose:

- the use of special thermostable frying fats containing food additives - antioxidants (inhibitors, synergists, complexing agents and antifoam agents) (TU 9142-018-00365517-2006);
- improvement of equipment using superheated steam to reduce the contact of fat with oxygen and the absorption of deep fat (Primo-Martin, C., 2011);
- maximum reduction in contact time of fat with the fried semi-finished product;
- reducing the duration of use of fat, replacing it with fresh one;
- the use of purification of used frying fat with active adsorbents (Yates R.A., 1992; Cooke B.S., 2006) followed by mixing refined and fresh deep frying fat and reusing the resulting fat mixture in the production cycle.

One of the most effective technological solutions to this problem is to reduce the absorption of frying fat and products of its thermo-oxidative degradation by finished products through the use of food additives (hydrocolloids) allowed by the technological regulations of TP CU 029/2012 (Vaskina V.A., 2014; Lvovich N.A., 2011; Varela P., 2011; Rogozin I.P., 2019). However, the use of these additives provides the necessary technological and economic effect only when the decrease in fat absorption by finished products reaches at least 20% relative to the content of deep frying fat in similar products produced without the use of food additives.

Food industry enterprises use mainly rapid tests to assess the suitability of fat for further use, primarily, changes in dielectric constant, increase in viscosity and density of deep frying fat. According to Marmesat et al. (Marmesat S., 2007; Marmesat S., 2008) these indicators are correlated with the degree of fat degradation.

German Society for Fat Science (DGF) indicates that used deep-frying fat should be considered unsuitable if it has an unacceptable taste and smell; in the case of satisfactory organoleptic characteristics. Fat is unsuitable if the content of fatty acid copolymers insoluble in petroleum ether is 0.7% or higher and at the same time smoke formation temperature is lower than 170 °C, or if the concentration of fatty acid copolymers insoluble petroleum ethers is higher than 1% (Fette, Seifen, Anstrichmittel, 75: 449, 1973).

In a number of foreign countries, deep frying fat is considered unsuitable for use when the content of the polar products of fat decomposition is 25%. In some European countries, this indicator is adopted by law.

As for the Russian Federation, the studies conducted at the Institute of Nutrition of the Academy of Medical Sciences of the USSR in 1960 (BrentsM. Ya., 1965; Botsman N.E., 1969), showed a close correlation between the content of fat oxidation products insoluble in petroleum ether and the effect of thermally oxidized fats on the body.

Table 1. Standardized indicators in the Russian Federation characterizing the level of deep frying fats safety

Indicator	Prescribedscale	Regulatorydocument
1. Content of thermostable oxidation products insoluble in petroleum ether, %	Notmorethan 1.0	SanPiN 2.3.6.1079-01
2. Peroxidecontent, meq/kg	Notmorethan 1.0	TR CU 024/2011
3. The content of FFA, mg KOH/g	Notmorethan0.6	TR CU 024/2011

The normative documents of the Russian Federation normalized the maximum content of thermostable oxidation products insoluble in petroleum ether in the deep frying fat. It is not more than 1% of the fat mass (Molchanova N.A., 1980).

Table 1 presents the standardized indicators in the Russian Federation characterizing the level of deep frying fats safety.

Along with secondary oxidation products, hydrolysis products, free fatty acids (FFA) and primary oxidation products (peroxides) are controlled in deep frying fats using methods for determining the acid and peroxide numbers of fats.

As the analysis of normative documentation showed, there are a number of different methods in the Russian Federation. But none of them, except for the laborious and lengthy method of determining the secondary oxidation products insoluble in petroleum ether, does not adequately assess the safety of deep frying fat.

On January 1, 2016, GOST R 54607.3-2014: Public catering services. Methods of laboratory quality control of products catering. Part 3. Methods of control of manufacturing processes of products of public catering", is carried out. Accordance to this document the quality control of deep frying fats is carried out in the following order.

The evaluation of the quality of deep frying fat used in the technological process begins with organoleptic indicators, in accordance with the rating scale and taking into account the importance factors. With an "unsatisfactory" organoleptic evaluation of deep frying fat, it must be replaced with fresh one. With a "satisfactory" organoleptic evaluation of deep frying fat, the degree of thermal oxidation should be determined according to this GOST.

The most important methods of instrumental quality control of deep frying fat in the technological process according to GOST R 54607.3-2014:

- spectrophotometric method for determining the degree of thermal oxidation of deep frying fat according to the content of conjugated dienes;
- methods for the determination of hydrolytic deterioration of deep frying fat by determining the acid number according to GOST R 52110 and GOST 504579 and using indicator test strips, taking into account recommendations for their use;
- method for determining the total content of polar substances in deep frying fat by measuring the dielectric constant using the Ebro and FOM-320 devices.

An analysis of the proposed deep frying fat quality control methods shows their wide range, but does not reveal the degree of their adequacy. It is not indicated which results of the determinations are critical for assessing the degree of thermal oxidation of deep frying fat.
GOST R 54607.3-2014 does not include some important methods:

- determination of the content of thermostable oxidation products insoluble in petroleum ether in deep fat
- method for determining the degree of thermal oxidation of deep frying fat from the reaction of dicarbonyl compounds, formed in the process of thermal oxidation of deep frying fats, with a solution of alcohol alkali, provided by the Guidelines for Laboratory Control of Food Products Quality (1991).

It was established that the color intensity of the alcohol-alkaline solution of thermally oxidized fats is proportional to the total content of the secondary thermostable products of oxidation and copolymerization calculated by the optical density. The amount of these products in deep frying fats should not exceed 1%.

The method for determining secondary oxidation products is laborious and requires a laboratory at a catering facility; therefore, the quality of deep frying fats is assessed using the above correlation methods. All these methods do not determine the secondary thermostable products of fat oxidation and copolymerization themselves, but other hydrolysis and oxidation products associated with correlation dependences with the secondary products of fat oxidation. The degree of correlation varies depending on the fatty acid composition of the deep frying fat, on the process conditions, and on the type of product to be fried. Therefore, the above methods are not effective at the assessment of the safety of wasted deep frying fat.

The complex composition of the products of hydrolytic and oxidative decomposition of deep frying fats makes it difficult to choose the most informative and adequate control indicators. Unfortunately, the methods currently used are suitable for determining the quality of deep frying fat, but not for determining its safety. These methods equate the determination of hydrolytic spoilage products and polar substances to the determination of thermal oxidation products.

In our opinion, the basis for assessing the suitability of deep frying fat for future use should be consumer safety criteria. The level of these indicators should be correlated with their impact on health.

To assess the real safety of fast food products and deep frying fats, it is necessary to control the intake of toxic substances into the body with the finished product. However, to date, this is not provided for by the current regulatory documents of the Russian Federation, which is the main drawback of the existing production control scheme.

Therefore, the development of a production control scheme for the safety of deep frying fats and fast food products and clarification of the acceptable level of toxic substances in these products is one of the most important tasks of this work.

Lipid oxidation is one of the main processes that limit the shelf life of food products in terms of safety and quality. However, to date, the safety control and shelf life of fast food products of industrial production are limited only by microbiological indicators without taking into account the safety of the fat component, the degree of change of which remains poorly understood in the technological process and at the stage of product circulation.

Given the sharply negative results of histological and hematological studies in experimental animals, given below in chapter 6 of this work, it is necessary to provide for full toxicological studies of oxidized deep frying fat and deep-fried finished products. This refers to the definition of the acceptable level of toxicants in the finished product, and of the allowable daily intake of toxic compounds, as well as the need of possible limitation of the daily intake of deep-fried products by the most labile consumer groups - children and students of secondary schools.

The results are of serious concern and require work to refine the list of indicators and strengthen production control of safety indicators of deep frying fat and deep-fried products.

The concretization of the processes occurring at high operating temperatures will allow:

- evaluating the intensity of oxidative changes in fats containing polyunsaturated fatty acids at temperatures of 170 - 180 °C;
- developing measures to protect deep frying fats from oxidation;
- evaluating existing methods for monitoring the safety of deep frying fats and deep-fried products;
- proposing measures to refine methods for monitoring.

The objectives of this study: a deeper study of the kinetics of hydrolysis and high-temperature oxidation of deep frying fats, clarification of the effect of specific fat decomposition products on product safety, improvement of toxic compounds control methods directly in "fast food" products.

5.2. The Effect of The Intensity of The Hydrolysis of Triacylglycerols On the Accumulation of Toxic Products of Fat Decomposition During Food Frying

The initial stage of research was aimed at elucidating the relative influence of the hydrolysis and thermal oxidation of deep frying fat on the deterioration of product safety.

The hydrolysis rate was determined by the value of the acid number (GOST R 52110-2003), at which the deep frying fat is replaced with fresh one. According to the current regulatory documentation, this is done when 1% of thermostable secondary oxidation products insoluble in petroleum ether are accumulated in deep fat oil.

In 2004, Federal Service for Surveillance on Consumer Rights Protection and Human Wellbeing has developed the "Guidelines for Determining the Quality of Deep Frying Fat with the 3M LRSM Indicator Strips (indicator of low concentrations of free fatty acids)". In accordance with these recommendations, deep frying fat is considered unsuitable for further use with an FFA concentration of more than 2 mg KOH/g of fat (approximately equivalent to 1.06% of FFA by weight of fat referred to oleic acid) (Recommendations, 2004).

The study of the hydrolysis of deep frying fat in St. Petersburg fast food chains McDonalds, Burger King, KFC showed that when the fat in fatty fat accumulates up to 1% of thermostable secondary oxidation products insoluble in petroleum ether, the acid number of deep fat was changed from 0.49 to 1.6 mg KOH/g, preferably from 0.5 to 1.0 mg KOH/g.

Although the kinetics of the accumulation of free fatty acids (FFA) correlates in time with the accumulation of thermostable secondary oxidation products insoluble in petroleum ether in deep frying fat, there is no justified causal relationship between these indicators. A single norm of acid number cannot be established at which the used deep frying fat should be replaced with fresh one.As a rule, over the entire period of use of deep frying fat, the acid number is relatively low. For this reason, the previously approved recommendations for the control of deep frying fat using indicator strips (GOST R 54607.3-2014) are practically inapplicable, since the acid number is low even after the normalized accumulation of thermostable secondary oxidation products insoluble in petroleum ether.

The value of the acid number of the finished product largely depends on the moisture content in the initial semi-finished product and the production technology. It is likely that each company can set its own standards for the acid number in the production of certain products, but it cannot be set to a single standard for the acid number, which should replace the used deep frying fat with fresh one.

To clarify the chemical composition of the resulting free fatty acids, they were isolated from oxidized deep frying fat on anionite AB-17 in carbonate form and converted to methyl esters directly on the anion exchanger (Golovkin N. A., 1970). Using gas-liquid chromatography (GOST R 51483-99), it was found that the isolated FFAs contain mainly compounds with a chain length of more than 12 carbon atoms.

The composition of FFA approximately corresponds to the fatty acid composition of fat, taking into account the more intensive hydrolysis of ester bonds in the α-positions of triacylglycerol molecules. FFA with a short chain of 3–9 carbon atoms was not detected, which indicates the formation of FFA solely as a result of hydrolysis (Perkel R.L., 2013).

Thus, in the early stages of deep frying fat destruction, polar products are formed mainly as a result of the hydrolysis of triacylglycerols and include monoacylglycerols, diacylglycerols and FFAs.

Some data on the formation of polar compounds during frying foods were obtained in the works of the Dallas Group experts (Yates R.A., 1992). In these works, it was shown that at the initial stages of the high-temperature process, most of the polar compounds are mono- and diglycerides and FFA (the products of hydrolytic changes in fat).

Therefore, in the initial stages of the process of thermal oxidation of deep frying fat, the oxidative transformations of deep frying fat do not yet lead to the formation of a noticeable amount of polar compounds.

Based on these theoretical principles, the presence of a correlation between the acid number of deep frying fat and the accumulation of polar compounds in it was verified. The determination of polar compounds was carried out using an Ebro instrument (Germany).

The experimentally established dependence of the accumulation of polar compounds on the acid number of deep frying fat is shown in Figure 1.

Figure 1. Dependence between the acid number of deep frying fat and the content of polar products in it (Personal research data)

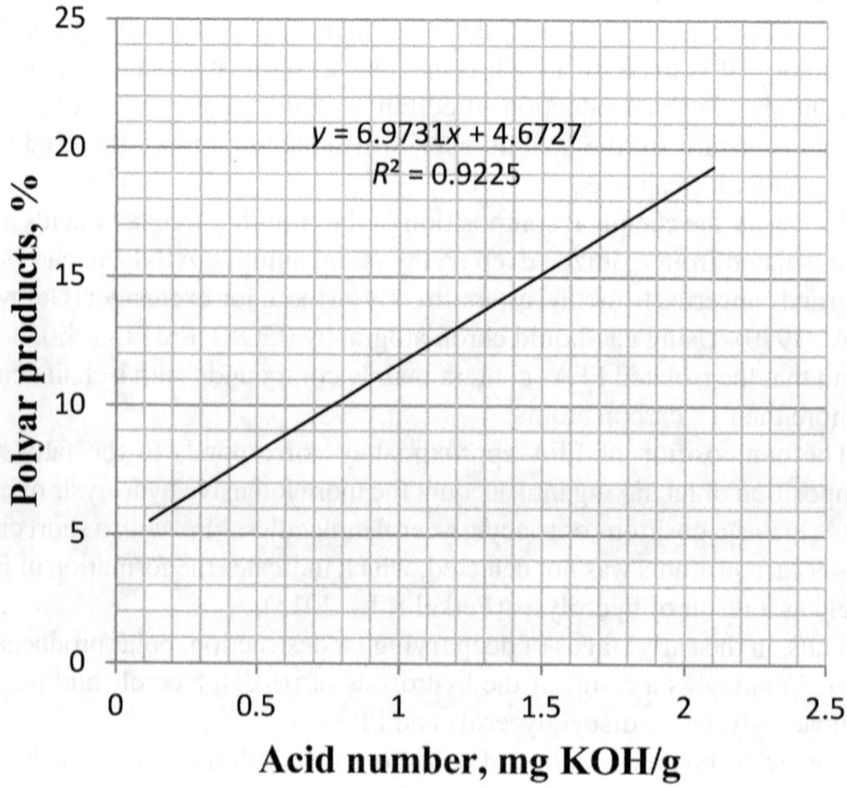

From the obtained data it follows that the definition of polar products and the value of the acid number give the wrong guidelines for replacing the deep frying fat. For example, at KFC, when the rate of polar products to replace the deep frying fat was 25.0%, the deep frying fat was actually replaced when the content of polar products was 17.0%.

Our studies have shown that at this enterprise the content of normalized secondary oxidation products insoluble in petroleum ether was 1.0% with the content of polar products of 15.0%.

Thus, recommendations for determining the content of polar products to control the degree of oxidation of deep frying fat also do not have sufficient scientific basis.

It should be noted in conclusion that FFAs, mono- and diglycerides are normal products of food digestion in the body and do not have any toxicity.

A real threat to human health is the products of high-temperature oxidation of deep frying fats.

5.3. Features of The Oxidation of Vegetable Oils During Deep-Frying

During deep-frying, the product absorbs a significant amount of fat (from 10 to 40%). Its safety for the consumer is largely determined by the safety of the absorbed fat, and it, in turn, is determined by the depth of the oxidative changes in fat in the technological process.Thermooxidative degradation results in the formation of compounds with hydroperoxide, epoxy, hydroxyl, carbonyl and carboxyl functional groups. Fat oxidation products cause serious changes in the biochemical composition and blood cells, lead to disruptions in antioxidant defense systems (Okunevich I.V., 2004; Simakova I.V., 2009; Makarova A.N., 2011).

The chemistry of the processes of radical chain oxidation of fats at moderate temperatures has been fairly well studied (Denisov E.T., 2005; Porter N.A., 1995; Girotti A.W., 2013). To a lesser extent, the features of oxidative transformations of unsaturated fats in deep-frying processes, characterized by elevated temperatures (150-180 °C), the presence of natural and synthetic antioxidants, the possibility of contact of fat with metal ions of variable valence, and some other factors are revealed.

5.3.1 Oxidation of polyunsaturated fats at moderate temperatures

According to modern concepts, the oxidation of fats by atmospheric oxygen proceeds according to a radical chain mechanism and includes the stages of chain initiation (1), chain growth (2 and 3), degenerate chain branching (4-6) and chain breaking (7-12). In these equations, the expression LH denotes polyunsaturated linoleic acid.

Chain initiation.

$LH + \bullet O\text{-}O\bullet \rightarrow L\bullet + HO_2\bullet$ (1a)

$LH + \bullet OH \rightarrow L\bullet + HO_2$ (1b)

$LH + RO\bullet \rightarrow L\bullet + ROH$ (1c)

Chain growth.

$L\bullet + \bullet O\text{-}O\bullet \rightarrow LOO\bullet$ (2)

$LOO\bullet + L_1H \rightarrow LOOH + L_1\bullet$ (3)

Degenerate chain branching and autocatalytic chain acceleration.

LOOH → LO• + •OH (4)

LO• + L_1H → LOH + L_1• (5)

•OH + L_2H → H_2O + L_2• (6)

Chain breaking by antioxidants.

LOO• +InH → LOOH + In• (7)

LOO• + In• → LOOIn (8)

Chain breaking at recombination of free radicals.

L• + L• → L- L (9)

L• + LO• → LOL (10)

LO• + LO• → LOOL (11)

LOO• + LOO• → LOOL + O_2 (12)

The speed of the process, the chemical composition and the quantitative accumulation of oxidation products differ significantly depending on the oxidation temperature from 20 to 180 ° C, on the content of polyunsaturated fatty acids in fat, the presence of natural and synthetic antioxidants and oxidation initiators.

The initiation of the oxidation chain occurs during the interaction of glycerides containing polyunsaturated fatty acids with free radicals contained in fat in trace amounts, as well as with activated oxygen molecules that become active as a result of sunlight influence.

The decisive influence is exerted by the natural spatial structure of polyunsaturated fatty acids, in which between the two double bonds in the carbon chain there is a methylene (-CH_2-) group:

-CH=CH-CH_2-CH=CH-

In chemistry, such a structure is called pentadiene (or divinylmethane) and differs in that, as a result of the influence of two neighboring double bonds, the hydrogen atom of the methylene group is activated, and the energy of abstraction of the hydrogen atom from the carbon atom of this group is not more than 335 kJ/mol.

Free radicals or photoactivated oxygen abstract a hydrogen atom from the active methylene group in the polyunsaturated fatty acids molecule, forming as a result of homolytic dissociation of the –CH- bond the free alkyl radical L• according to equations (1a – 1c). It is the first stage of the oxidation chain reaction.

The reaction of initiation of oxidation chains in a real technological process is very easy and practically does not require photoactivation of oxygen, since, the content of lipid hydroperoxides LOOH in food is from 1 to 100 µmol per 1 kg of fat. During transportation and storage, the content of hydroperoxides in culinary and deep frying fats of industrial production increases even more and reaches 1-3 mmol of active oxygen / kg of fat.

The energy of homolytic dissociation of the -O-O- bond in LOOH hydroperoxides for two radicals is comparatively low and amounts to approximately 151 kJ/mol. Under the influence of high heat treatment temperatures and in the presence of prooxidants, hydroperoxides undergo homolytic decomposition of the –O-O- bond according to Eq. (4) with the formation of the hydroxyl radical •OH and the alkoxyl radical LO• (the active initiators of the oxidation of unsaturated fatty acids).

The reaction of the breaking of hydroperoxides into radicals is catalytically accelerated by metal ions of variable valency, in particular, by iron and copper:

$LOOH + Fe^{+2} \rightarrow Fe^{+3} + LO\bullet + OH^-$ (13)

$LOOH + Fe^{+3} \rightarrow Fe^{+2} + LOO\bullet + H^+$ (14)

Particularly active catalysts for the hydroperoxides breaking are hemoprotein components - the breakdown products of hemoglobin, myoglobin, cytochrome C, which enter the deep frying fat when frying products of animal origin.

The formation of the radical L• according to equations (1a - 1c) is the easier, the lower the strength of C-H bond. During the oxidation of linoleic acid the hydrogen atom easily detaches from the active methylene group at the 11th carbon atom located between two double bonds. In this case, the pentadienyl (bis-allyl) radical L• is formed, characterized by a uniform distribution of electron density between all five carbon atoms (from the 9th to the 13th).

Oxygen dissolved in fats attaches to the L• radical, forming the less active hydroperoxyl radical LOO• (equation 2). The addition of an oxygen molecule to the pentadienyl radical of linoleic acid occurs at the 9th or 13th carbon atoms. In this case, isomeric hydroperoxyl radicals 9-LOO• and 13-LOO• are formed with a system of conjugated cis-trans- or trans-trans-double bonds (conjugated dienes). The higher the oxidation temperature, the more conjugated dienes are formed in trans-trans form.

Thus, as a result of reaction (2), oxygen is incorporated into the structure of polyunsaturated fatty acids:

Further, the formed hydroperoxyl radical LOO• detaches the hydrogen atom from another molecule of the polyunsaturated fatty acid L_1H by reaction (3), forming the hydroperoxide LOOH and the new alkyl radical L1•.

It is this reaction (3) that detaches a hydrogen atom from an oxidizable fatty acid that is limiting, since its rate constant is lower than the reaction rate constant (2). The chain free-radical reaction of fat oxidation is carried out by alternating these reactions.

Active methylene groups of polyunsaturated fatty acids enter the oxidation reaction most easily. There are no such groups in the oleic acid molecule, in the molecules of linoleic, linolenic and arachidonic acids, there are one, two and three active methylene groups, respectively. It was found that the oxidation rate of these acids under comparable conditions is 1: 12: 25: 50.

At the initial stage of oxidation, the concentration of radicals is low and constant, and the concentration of reaction products (hydroperoxides) increases linearly.

With the accumulation of hydroperoxides in deep frying fat, their homolytic decomposition into radicals increases according to reaction (4), especially in the presence of metals of variable valency. This reaction, called degenerate chain branching, leads to the fact that as a result of one reaction cycle (equations 2 - 4) three new free radicals (L_1•, LO•, •OH) are formed from one radical L•. Each of these free radicals is able to initiate new oxidation chain. The total concentration of radicals increases sharply.

The accelerated increase in radical concentration leads to a very rapid accumulation of reaction products (hydroperoxides). The reaction is self-accelerating, that is, it is an autocatalytic reaction (equations 4-6).

Hydroperoxides are the primary oxidation products. They do not have a taste or smell, but are harmful to the body, as they are able to interact with proteins and lipids, disrupting the work of a number of enzyme systems and cell membranes. The content of hydroperoxides in food fats is allowed no more than 10 mmol/kg of active oxygen.

The system of conjugated double bonds is retained in the LO• radical, in the LOH hydroxyacids formed from it, as well as in some products of the destruction of the LO• radical, for example, in α-, β-unsaturated aldehydes.

The destruction of the LO• radical by the β-cleavage reaction proceeds with the breaking of the carbon chain and the formation of various secondary oxidation products (free aldehydes, hydrocarbons, free fatty acids and triglycerides containing oxidized fatty acid chains in their structure). These secondary oxidation products with a chain length of 3 ... 12 carbon atoms have an intense unpleasant taste and aroma.

The accumulation of volatile compounds with an average chain length (carbonyl compounds, free fatty acids) leads to a deterioration in the organoleptic characteristics of edible fats. There is evidence that a marked deterioration in the organoleptic characteristics of fat occurs when 4-5% triacylglycerols are oxidized and up to 100 meq O_2/kg of fat is absorbed (Ma`rques-Ruiz G., 1996).

The limited access of oxygen during deep fat frying promotes the intensification of chain breaking by recombination of radicals with the formation of non-polar and low-polar oxygen-containing polymers according to reactions (9) - (12) in the early stages of the oxidation process.

Thus, the following factors influence the intensity of radical chain oxidation of fats at moderate temperatures:

- content of polyunsaturated fatty acids (active methylene groups);
- chain initiation by traces of free radicals, photoactivated oxygen and decomposition products of hematoprotein components containing ferric iron;
- chain development by the reaction of radicals with molecular oxygen;
- the breaking of hydroperoxides into radicals, catalyzed by iron, copper iones, or by hemoprotein components.

5.3.2 Features of High-Temperature Oxidation of Deep Frying Fats

The rate and direction of oxidation reactions of unsaturated fatty acids vary significantly at deep-frying temperatures (150-180 °C).

With increasing temperature, the oxidizing properties of the LOO• peroxyl radical are enhanced. At deep-frying temperatures, LOO• peroxyl radicals can not only detach hydrogen from the LH substrate by reaction (3), but also oxidize double bonds to epoxy groups by reaction (15) (Denisov E.T., 1983; Dryuk V.G., 1999; Guiffrida F., 2004):

It is known that epoxy compounds are able to form during the oxidation of double bonds with hydroperoxides, especially in the presence of catalysts (Dryuk V.G., 1999):

Reaction (15) involves the parallel formation of compounds with hydroperoxide and epoxy groups, while with the predominance of reaction (16), epoxy compounds will be formed sequentially with hydroperoxides, which may affect the kinetics of epoxides accumulation.

Due to spatial obstacles, the LOO• radical is unable to oxidize conjugated double bonds in its own chain, but only in the adjacent chain of unsaturated fatty acid. This mechanism explains the location of the epoxy cycle at the double bond site and the simultaneous formation of hydroxy acid with conjugated double bonds from the initial hydroperoxide.

The possibility of the formation of epoxides during the oxidation of unsaturated compounds in biological systems and in mixtures containing unsaturated hydrocarbons was established in the works of Russian scientists (Emanuel N.M., 1961; 1965; 1966; Denisov E.T., 1983). The late formation of the epoxy cycle by the reaction of the LOO• hydroperoxyl radical with a double bond in the external fatty acid chain was confirmed in (Guiffrida F., 2004).

During the oxidation of linoleic acid, this reaction produces 9,10-epoxy-12-cis-octadecenoic acid and its transisomer, as well as 12,13-epoxyoleic acid or its trans-isomer. It was established that the first isomer is leukotoxin, and the second isomer is isoleukotoxin. In an animal experiment, these toxins provoke a decrease in the white blood cell count and exhibit carcinogenic activity (Greene J.F., 2000; Wilson R., 2002; Markaverich B.M., 2005). Toxic epoxyoctadecenic acids represent a significant proportion of the products of thermal oxidation of olive and sunflower oils (up to 3%) (Velasco J., 2004).

Peroxyl radicals and peroxides are capable of oxidizing a double bond not only in polyunsaturated fatty acids, but also in monounsaturated fatty acids, for example in oleic acid. In this case, derivatives of epoxystearic acid (cis- and trans-isomers) are formed that do not contain a double bond (Velasco J., 2004).

At the later stages of thermal oxidation of linoleic acid, diepoxy acids may form, and during the destruction of the LO• radical, carbonyl compounds containing oxy- and epoxy groups in the chain and having significant toxicity are formed (Guillen M.D., 2008; Goicoechea E., 2010).

Thus, the processes of thermal oxidation of deep frying fat significantly differ from the usual process of autooxidation of fats due to a significant increase in the oxidative activity of the LOO• radical at high temperatures.

At elevated temperatures, the accumulation of the most polar components (dihydroxy acids insoluble in petroleum ether) significantly increases.

The accumulation of dihydroxy acids insoluble in petroleum ether in deep-fried finished products is highly undesirable. In the work of BrentsM. Ya.(BrentsM.Ya., 1965), previously performed at the Institute of Nutrition of the Academy of Medical Sciences of the USSR, it was found that the content of dihydroxy acids insoluble in petroleum ether in oxidized edible fats should not exceed 1% by weight of fat. In modern biological studies in animals, performed in the Saratov State Agrarian University named after N.I. Vavilov (Simakova I.V., 2009; Makarova A.N., 2011), it has been shown that dihydroxy acids insoluble in petroleum ether have a negative effect on the organs of the gastrointestinal tract, cardiovascular system, biochemical composition and blood formula of experimental animals (PFB Chapter 7).

However, in the available scientific literature, we did not find data that allowed us to estimate the rate of accumulation of epoxides and conjugated dienes as a result of oxidation of double bonds by peroxyl radicals and peroxides depending on the oxidation temperature, the content of polyunsaturated fatty acids in deep frying fat, the presence of antioxidants, and other process features.

5.3.3 The Study of The Kinetics of Accumulation of Products of Thermal Oxidation of Unsaturated Fats During Deep Frying

The purpose of this section of the work was to study the features of the oxidation of deep frying fats of industrial production used in fast-food chains of St. Petersburg, and their impact on consumer properties of products. The studies were conducted at St. Petersburg State University of Trade and Economics by Volovey A.G. under our guidance from 2012 to 2015 (Perkel R.L., 2013, 2014).

For research we used Vegafray brand fats according to TU 9142-018-00365517-2006 “Special Purpose Fats” manufactured by Cargill LLC, containing from 9.4 to 66.0% polyunsaturated fatty acids (PUFA). Vegafray 05 and Sunny Gold fats contained antioxidant food additives authorized by TS TU 029/2012: oxidation inhibitor E321 (Butylhydroxytoluenum), complexing agent E330 (citric acid), silicone defoamer E900 (polydimethylsiloxane).

The study of thermal oxidation of deep frying fat in the presence of the product was carried out under production conditions, in an electric deep fat fryer with a capacity of 7.0 dm^3 of oil, during the manufacture of French fries at a temperature of 180 °C from partially fried semi-finished product. In the intervals between the production of individual batches, the temperature of the fat was not lower than 150 °C.

Potato was fried in cycles of 12-16 hours; in between, the fat was allowed to cool at a temperature of 20 °C for 8-12 hours. Model fatty fat oxidation without a product was carried out in a thin layer in thermostated Petri dishes with periodic gentle stirring (Perkel R.L., 2014).

Thus, the process of fat oxidation took place in a diffusion mode, both in the presence of the product and without it.

The fatty acid composition of the fats was determined by gas-liquid chromatography of methyl esters of fatty acids (GOST R 51483-99, GOST R 51486-99). The peroxide and acid number of the initial and oxidized fats was determined by standardized methods (GOST R 51487-99, GOST R 52110-2003).

The actual content of secondary thermostable oxidation products insoluble in petroleum ether was determined according to the method described in the VNIIZh Manual (Manual, 1967a), in the author's modification (Perkel R.L., 2014).

A mixture of oxidation products insoluble in petroleum ether was dissolved in hot ethanol, concentrated in a heated bath to a small volume, transferred quantitatively to a 50 cm^3 volumetric flask, made up to the mark with ethyl alcohol and mixed well.

The resulting solution containing secondary thermostable oxidation products insoluble in petroleum ether was divided into 2 equal parts. One part was dried to constant weight, and then calcined to determine ash, as described in the basic procedure. According to the data obtained, the mass fraction of secondary thermostable oxidation products insoluble in petroleum ether in oxidized fat was calculated, taking into account that the treated solution contains secondary thermostable oxidation products insoluble in petroleum ether from 2.5 g of fat.

In the second part of the solution, the fatty acid content was determined by titration with a 0.1 mol/dm^3 alkali solution. According to the results obtained, the content of secondary thermostable oxidation products insoluble in petroleum ether in oxidized fat in mmol/kg of fat and the average molecular weight of secondary thermostable oxidation products insoluble in petroleum ether were calculated.

The content of epoxides was determined by the USSR author's certificate No. 1040914 (Stopsky V.S., 1983). The analyzed fat sample was treated with 82-87% phosphoric acid, followed by precipitation of the unreacted acid with a non-polar hydrocarbon solvent (hexane). After removing the precipitate of unreacted acid, the amount of phosphoric acid that reacted with epoxides was determined by centrifugation and removing the solvent under vacuum in the analyzed sample, and the amount of epoxides in fat was calculated.

It has also been developed a methodology for the determination of epoxides by FTIR-spectroscopy (Tavassoli-Kafrani M.H. et al., 2017; Nzai J.M. et al., 1998) by the number of oxyphosphates formed under the action of phosphoric acid.

The determination of secondary oxidation products containing conjugated double bonds was carried out according to GOST 54607.3-2014. The optical density of the fat solution in hexane "*D*" was determined at a wavelength of 232 nm in a 10 mm thick cuvette using an SF-26 spectrophotometer.The test solution was prepared by dissolving 0.05–0.1 g fat sample in a volumetric flask with a capacity of 100 cm^3. According to the measurement results, the specific absorption "*E*" was calculated by the following formula:

E = D/P, (18)

where *P* is a test fat sample, g.

With a specific absorption of less than 15, deep frying fat is considered suitable for further use, that is, the content of thermostable secondary oxidation products (TSOP) in it does not exceed 1%. Based on the results obtained, the calculated percentage of TSOP in the test fat was determined by the following formula:

TSOP = E/15, . % (19)

To express the content of conjugated dienes in mmol/kg, the specific absorption "*E*" was recalculated using a molar repayment ratio of 26040, calculated according to the data of the All-Russia Scientific Research Institute of Fats ("Manual", 1967b).

Fat was extracted from the finished product by the extraction-weight method according to GOST R 54607.8-2016.

In accordance with the task at the first stage of research, the oxidation of Sunny Gold deep frying fat was studied during heating without a product to assess the most important differences in the thermal oxidative degradation of unsaturated fatty acids at relatively low and high temperatures (Perkel R., 2014).

Oxidation was carried out in a thin layer of fat, in Petri dishes, at temperatures of 50 and 180 °C. The temperature of 50 °C was chosen because the thermal decomposition of hydroperoxides at this temperature occurs with low intensity. Kinetic curves of the accumulation of peroxides and epoxy compounds are shown in Figures 2 and 3.

Figure 2 shows that at 50 °C there is a small induction period of peroxide accumulation (about 2 days), after which the concentration of peroxides increases almost linearly, at a constant rate. Thus, natural and synthetic antioxidants in deep frying fat actually protect fat only during transportation and storage, before it is used in the process. The accumulation of peroxides at a constant rate, without auto-acceleration, indicates the almost complete absence of degenerate branching reactions at this temperature.

Epoxy compounds at 50 °C are formed in small quantities, their concentration slightly increases from 2.7 to about 7.5 mmol/kg (Figure 2). Preserving the rate of accumulation of peroxides and accelerating the accumulation of epoxy compounds indicate the formation of epoxides by reaction (15).

Figure 2. Kinetics of the accumulation of peroxides (1) and epoxy compounds (2) during the oxidation of Sunny Gold deep frying fat at 50 °C (Personal research data)

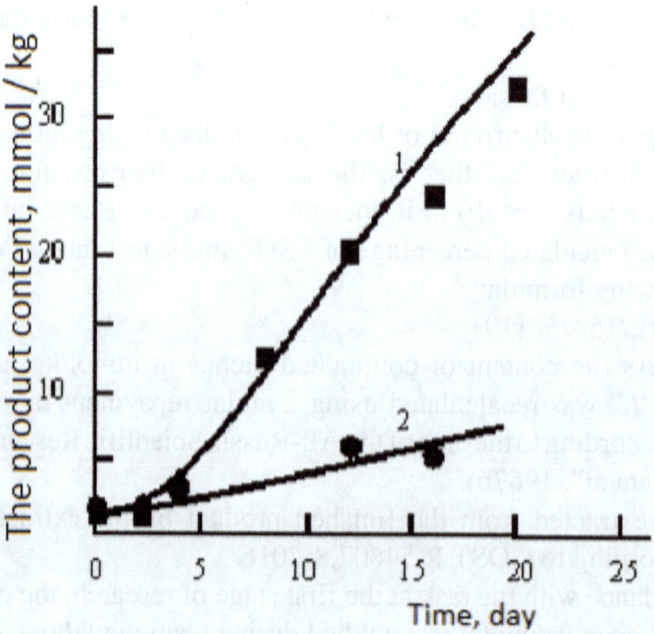

Figure 3. Kinetics of the accumulation of peroxides (1) and epoxy compounds (2) during the oxidation of Sunny Gold deep frying fat at 180 °C (Personal research data)

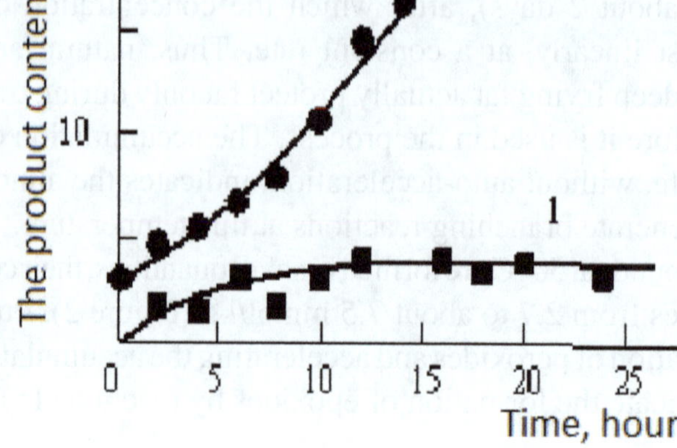

The kinetic picture changes upon transition from a temperature of 50 °C to a temperature of 180 °C (Figure 3).

At 180 ° C, the hydroperoxide accumulation curve reaches a low stationary concentration of 1-3 meq O_2/kg. The latter circumstance is characteristic of intermediate products and indicates that peroxides under experimental conditions interact with double bonds of unsaturated fatty acids with high rate.

At the same time, a significantly faster increase in the content of epoxy compounds to values of 24-28 mmol/kg is observed. It can be assumed that this is mainly associated with an increase in the overall oxidation rate at increasing temperature from 50 to 180 °C.

The presence of slight acceleration in the initial section of the kinetic curve of the accumulation of epoxy compounds suggests their partial formation by reaction (16) with the participation of hydroperoxides, and the characteristic S-shaped curve indicates their further expenditure for the formation of polymer products with a more complex composition.

At the same time, the products of reactions (15) and (16) - epoxides - accumulate in fat.

It is important to compare the intensity of competing reactions of the LOO• radical: reaction (3), which leads to the formation of hydroperoxides, and reaction (15), which leads to the accumulation of epoxides.

According to the work (Denisov E.T., 1983), the quantitative ratio of the formed peroxides to epoxides by reactions (3) and (15) during the autooxidation of unsaturated compounds under ordinary conditions (20–25 °C)

Table 2. The kinetics of accumulation of oxidation products of Sunny Gold deep frying fatdepending on the duration of potatoes frying (Personal research data)

The duration of the process, h	Fat hydrolysis and oxidation rates				
	Peroxide value, meq O_2/kg	**Acid number, mg KOH/g**	**Conjugateddienes, mmol/kg**	**Epoxides, mmol/kg**	**Weight content of copolymers insoluble in petroleum ether, %**
0	2.88	0.11	7.1	4.9	0
4	1.13	0.18	35.7	5.4	0.33
8	1.17	0.27	53.9	11.5	0.47
12	0.63	0.35	-	13.3	0.64
16	1.82	0.44	60.0	14.6	0.69
20	2.24	0.29	58.2	21.7	0.75
24	1.1	0.30	62.9	26.4	0.79
28	2.0	0.43	-	27.0	0.87
30	2.0	0.49	-	28.8	1.11

is approximately 10: 1. According to the above data from our studies, when the frying fat was oxidized at 50 °C, the ratio of the resulting hydroperoxides to epoxides was 32.5: 7.5 = 81: 19.

Studies characterizing the actual process of deep frying fat oxidation in the presence of the product were carried out at 180 °C.

The kinetics of accumulation of oxidation products of Sunny Gold deep frying fat in the technological process for the production of French fries from partially fried semi-finished products is given in Table 2 (Perkel, 2014).

As follows from the above data, after 28 hours of thermal oxidation, that is, after about 2 days of using Sunny Gold deep frying fat in the actual process of making French fries, the deep fat frying complies with the requirements of current regulatory documents for safety indicators.

Peroxide and acid number of deep frying fat met the requirements of TS TU 021/2011, and the weight content of copolymers insoluble in petroleum ether met the requirements of SanPiN 2.3.6. 1079-01 (not more than 1%).

The intensity of oxidation reactions at 180 °C was evaluated by the accumulation of conjugated dienes and epoxides. When the content of conjugated dienes was 62.9 mmol/kg of fat, the ratio of conjugated dienes to epoxides significantly decreased and amounted to 62.9: 26.4 = 70: 30.

Further studies of the kinetics of the accumulation of epoxy compounds during deep-frying of fruits and vegetables in batter and products from yeast dough (crumpets and donuts) were continued in experiments carried out by Rogozin I.P. under our guidance (Rogozin I.P., 2018a) at the technological laboratories of the Graduate School of Biotechnology and Food Science of Peter the Great St. Petersburg Polytechnic University.

When deep-fried dough pieces using "Reinbow" deep frying fat (Lithuania), the following data were obtained (Table 3).

Table 3. Weight content of conjugated dienes and epoxides in "Reinbow" deep frying fat (Lithuania) depending on the time of battered frying fruits (Personal research data)

Fryingtime, hour	Weight content of conjugated dienes, mmol/kg	Weight content of epoxides, mmol/kg	Weight content of copolymers insoluble in petroleum ether, %
0	21.5	6.0	0
2	28.7	31	0.47
4	38.0	54	0.69
6	54.8	-	0.79
8	60.9	64.5	1.11

As can be seen from the data presented, in the initial period, the content of epoxides is much lower than the content of conjugated dienes, since during storage and transportation of deep frying fat, oxidation occurs at moderate temperatures according to the auto-oxidation scheme. The ratio of conjugated dienes to epoxides is 21.5: 6.0 = 78:22.

However, with continued deep-frying, the content of epoxides exceeds the content of conjugated dienes; the ratio of conjugated dienes to epoxides approaches 50:50. After about 7.5 hours of deep-frying and at weight content of copolymers insoluble in petroleum ether at 1%, the weight content of conjugated dienes was about 59 mmol/kg, and the weight content of epoxides was about 63 mmol/kg.

Table 4. Weight content of conjugated dienes and epoxides during deep-frying of crumpets and donuts using the Sunny Gold deep frying fat (Personal research data)

Fryingtime, hour	Weight content of conjugated dienes, mmol/kg	Weight content of epoxides, mmol/kg	Weight content of copolymers insoluble in petroleum ether	
			%	mmol/kg
0	30.4	9.8	0.17	7.5
4	49.2	45	0.82	37
6	57.5	65	1.0	45
8	69.2	108	1.46	59

At the initial moment, the ratio of conjugated dienes to epoxides was 30.4:9.8 = = 76:24. After 4 hours of Sunny Gold deep frying fat use, the content of epoxides approached the content of conjugated dienes, and after 6 hours of deep-frying at the maximum permissible weight content of copolymers insoluble in petroleum ether of 1%, the weight content of conjugated dienes was 57.5 mmol/kg, weight content of epoxides 65 mmol/kg.

Among the products of the destruction of the LO• radical, it is necessary to specificate the most polar products - fatty dihydroxyacids containing two hydroxyl groups in the fatty acid chain. After fat saponification, these dihydroxyacids were isolated by a special technique in the form of a fraction of oxidized fatty acids insoluble in petroleum ether ("Manual", 1967a; Perkel R.L., 2014).

The developed original method for determining the molecular weight of copolymers insoluble in petroleum ether allowed us to prove that it is 222–225 g/mol; hence, 1% of copolymers insoluble in petroleum ether corresponds to

45 mmol/kg (Rogozin, 2018b). The expression of copolymers insoluble in petroleum ether in% of the mass of fat allows one to control the accumulation of copolymers insoluble in petroleum ether in deep frying fat and finished products. The expression of copolymers insoluble in petroleum ether in mmol/kg is useful for assessing the ratio of the products of oxidation and fat hydrolysis.

The study of the features of high-temperature oxidation of deep frying fats, performed in the works ofVolovey A.G. and Rogozin I.P., allows us to make the following conclusion:

1. The oxidation of deep frying fats in model experiments without a product and under production conditions in the presence of the product proceeds according to a radical-chain mechanism
2. The kinetics of thermal oxidation of deep frying fats at 150-180 °C differs significantly from the process of autooxidation due to the increased oxidizing ability of the LOO• radical and LOOH hydroperoxides. The concentration of peroxides remains approximately constant at a low level of 1 to 3 mmol/kg, which is typical for the intermediate products of the oxidation reaction. In this regard, the concentration of hydroperoxides in the product cannot be used to assess the degree of oxidation of deep frying fat.
3. Conjugated dienes, epoxides and highly polar dihydroxy acids insoluble in petroleum ether accumulate as a result of thermal oxidation in fat and in deep-fried foods.
4. at the maximum permissible weight content of copolymers insoluble in petroleum ether of 1%, the weight content of conjugated dienes was 57.5 mmol/kg, weight content of epoxides 65 mmol/kg.
5. The molecular weight of copolymers insoluble in petroleum ether is 222–225 g/mol; hence, 1% of copolymers insoluble in petroleum ether corresponds to 45 mmol/kg.

5.4. Studies of Safety Indicators Upon Storage of Fast Food Products with A High Content of Fat Component

The products of fast food network enterprises and of industrial production most popular with consumers were selected as the objects of research. They were produced by various manufacturers, using deep-frying and without frying, had a large share of the fat component, various price categories. The shelf life of the studied fast food products from the moment of production and time frame for the samples study are indicated in Table 5.

Table 5. Name, shelf life and time frame for study from the moment of production of fast food products (Personal research data)

Name and characteristics of fast food products	Shelflife	Time frame for study from the moment of production
National flour culinary product "Chak-chak" of industrial production	3 months	2 weeks
French fries of industrial production from fresh peeled chopped sliced potatoes		
Hamburgers from pre-frozen self-service meats of industrial production		
Deep-fried potato and corn chips of various manufacturers	7-9 months	3 months
Shortbread cookies of industrial production	6 months	3 months
Croutons of industrial production	6 months	3 months
Crackersof industrial production	6 months	3 months
Cereal bar with fruits for a healthy diet containing a fat component	6 months	3 months

The mass fraction of fat in culinary products and fast food products was determined by the extraction-weight method according to GOST 54607.8-2016. Fat was extracted from the product with diethyl ether. The solvent was removed from the extract under mild conditions, under vacuum, and the resulting fat was weighed to determine the mass fraction of the fat component.

The mass fraction of methyl esters of individual fatty acids to their total was determined by gas-liquid chromatography according to GOST R 51483-99 and GOST R 51486-99. The peroxide and acid number of the fat component was determined by standardized methods (GOST R 51487-99, GOST R 52110-2003). The total content of secondary oxidation products insoluble in petroleum ether was determined according to the VNIIZh methodology ("Manual", 1967a).

Studies of the mass fraction of fat showed that the fat content in products ranges from 8 to 28% (Table 6).

By the ratio of the main fatty acids in the fat component of the products, it was found that in the manufacture of these products vegetable oils and their mixtures, mainly palm and sunflower, were used as deep frying fat. It

Table 6. Mass fraction of fat in investigational product, % (Personal research data)

Product	Massfractionoffat, %
Fast Food Networking deep-frying product	
French fries, Russia	15.0
Friedchickenwings, Russia	22.0
Hamburgers from self-service meats, Russia	20.0
Deep-frying product of industrial production	
Potato chips in slices, Russia	27.8
Potato straw chips, Poland-Germany	26.9
Cornchips, Russia	17.9
Chak-chak, Russia	17.6- 31.0
Snack products, including snacks for a healthy diet	
Crackers, Germany	16.9
Croutons, Russia	8.5
Shortbreadcookies, Russia	19.7
Cerealbar, Slovakia	11
Cerealbar, CzechRepublic	8.6

should be noted that the fatty acid composition of the studied products does not comply with the FAO / WHO recommendations for a healthy diet due to the high content of saturated fatty acids.

The safety study of the fat component (Table 7) in the products was carried out at the stage of their implementation in fast food enterprises and in the retail trade network (after 3 months of storage).

Determination of the acid numbers of the studied products showed that the most intense accumulation of free fatty acid occurs in the fat component of croutons, corn chips, potato straw chips, crackers, and healthy nutrition bars. The indicators of the acid number of the fat base of potato chips in slices and shortbread cookiesare within normal limits.

The value of the acid number in stored products largely depends on the moisture content in the initial semi-finished product and the production technology, as well as on storage conditions and the initial safety parameters of the fat ingredient used in the formulation.

Analysis of the data in Table 7 showed that the highest content of hydroperoxides, exceeding the permissible norm, was in the fat component of potato chips in slices, corn chips, and of chak-chak, respectively. These

Table 7. Safety indicators of the fat component and finished fast food products (Personal research data)

Product	Acid number, mg KOH/g	Peroxide value, meq O_2/kg	Weight content of copolymers insoluble in petroleum ether, % by weight of the fat component	Weight content of copolymers insoluble in petroleum ether, % by weight of finished products
French fries	0.49	2.0	1.11	0.16
Friedchickenwings	0.77	17.1	1.25	0.27
Fried self-service meats	2.8	5.48	2.08	0.42
Potato chips in slices	0.39	10.5	2.55	0.7
Potato straw chips	1.0	2.9	1.0	0.3
Cornchips	0.8	8.8	1.3	0.23
Chak-chak, "Chelnykhleb", Tatarstan	2.7	13.6	1.6	0.3
Noodles"Rolton", Russia	0.4	2.4	1.4	0.3
Crackers "TUC", Germany	0.7	1.2	0.7	0.1
Croutons S-ZAO "Bridgetown Foods", Russia	0.79	6.6	1.32	0.1
Shortbread cookies «Aro", Russia	0.41	3.5	3.65	0.7
Cereal bar "Ego" with cherry, Slovakia	1.9	1.4	1.2	0.1
Cereal rich bar "Fit", Czech Republic	2.6	4.2	2.1	0.2

products have the largest surface area covered with fat, which is subjected to intense oxidation during storage.

The results of the determination of the content of compounds insoluble in petroleum ether in the studied extracted fats, presented in Table 7, indicate that their concentration in the fat component of most fast food products is 1.2-3.6 times higher than the norm established for deep frying fat, and ranges from 0.1 to 0.7% by weight of the product.

Results of the study indicate that there is a need to monitor the safety of finished fast food products and to review the timing of the implementation of these product groups, taking into account the safety indicators of the fat component.

FURTHER RESEARCH IN THIS AREA

The results are of serious concern and urgently require work to refine the list of indicators and strengthen production control of safety indicators of deep frying fat and deep fried products.

It is necessary to find out the effect degree of sopolymers insoluble in petroleum ether and epoxides on the determination of the safety of deep-fried products. Further research should be aimed at obtaining product samples containing individually sopolymers insoluble in petroleum ether and epoxides, and at assessing their impact on product safety for the body.

It is necessary to develop a method for purifying deep frying fats from epoxides.

Given the sharply negative results of histological and hematological studies in experimental animals, given below in chapter 6 of this work, it is necessary to provide for full toxicological studies of oxidized deep frying fat and deep-fried finished products. This refers to the definition of the acceptable level of toxicants in the finished product, and of the allowable daily intake of toxic compounds, as well as the need of possible limitation of the daily intake of deep-fried products by the most labile consumer groups - children and students of secondary schools.

REFERENCES

Adhikari, C. (1994). Diffuse-reflectance fourier-transform infrared spectroscopy of vegetable oil triglyceride adsorption on silicic acid. *J. Amer. Oil Chem. Soc., 17*(6), 589-594.

Botsman, N. E. (1969). The influence of repeatedly warmed sunflower oil on the histological structure. *Questions of a balanced diet (Kiev)*, (5), 32–37.

Brents, M. Y. (1965). *A study of changes in vegetable oils during the heat treatment of products in them at public catering establishments.* Ph.D. thesis in Engineering Science.

Cooke, B. S. (2006). Adsorbent Treatment of Frying Oil: Commercial Frying Case Study. Abstracts World Conference and Exhibition on Oilseed and Vegetable Oil Utilization, General Session 111: Frying Oils. *Oil Stability*.

Denisov, E. T. (1983). Oxidation and stabilization of rocket fuels. Academic Press.

Denisov, E. T. (2005). *Oxidation and Antioxidants in Organic Chemistry and Biology*. Taylor and Francis.

Dobarganes, M. C. (1987). Non polar dimer formation during thermoxidation of edible oils. *Fat Science and Technology.*, *89*, 216–220.

Dryuk, V. G. (1999). *Oxirans - synthesis and biological activity. The mechanism, associative and steric effects of peroxide epoxidation of alkenes*. M.A. Voytsekhovskaya.

Emanuel, N. M. (1961). *Inhibition of fat oxidation processes*. N.Lyaskovskaya.

Emanuel, N. M. (1965). *Chain reactions of oxidation of hydrocarbons in the liquid phase*. Z.K. Mayzus.

Girotti, A. W. (2013). Generation and Reactivity of Lipid Hydroperoxides in Biological Systems. In Patai's Chemistry of Functional Groups. John Wiley & Sons, Ltd. DOI: 0864 doi:10.1002/9780470682531.pat

Goicoechea, E. (2010). Analysis of hydroperoxydes, aldehydes and epoxydes by ^{1}H nuclear magnetic resonance in sunflower oil oxidized at 70 and 100 ºC. *Journal of Agricultural and Food Chemistry*, *58*, 6234–6245. doi:10.1021/jf1005337 PMID:20438132

Golovkin, N. A. (1970). Analysis of free fatty acids in natural fats by gas-liquid chromatography. Scientific Papers of VNIIZh, 27, 247–252.

GOST R 51483-99.GOST 51483-99 Vegetable oils and animal fats. Determination by gas chromatography of constituent contents of methyl esters of total fatty acid content. URL: vsegost.com.

GOST R 51486-99.Animal fats and vegetable oils.Preparation of methyl esters of fatty acids.URL: vsegost.com.

GOST R 51487-99.Vegetable oils and animal fats.Method for determination of peroxide value.URL: vsegost.com.

GOST R 52110-2003.Vegetable oils.Methods for determination of acid value. URL: vsegost.com.

GOST R 54053-2010. Methods for determination of fat weight fraction. URL: vsegost.com.

GOST R 54607.3-2014. Methods of laboratory control of catering products. Part 3.Methods for monitoring compliance with the processes of manufacturing catering products. 6. Quality control of deep frying fat, p. 2-9. URL: vsegost.com.

GOST R 54607.5-2015. Methods of laboratory control of catering products. Part 5.Methods for fat determination. URL: vsegost.com.

GOST R 54607.8-2016. Methods of laboratory control of catering products. Part 8. Accelerated control methods. URL: vsegost.com.

Greene, J. F. (2000). *Toxicity of epoxy acids and related compounds to cells expressing human soluble epoxide hydrolase* (Vol. 13). Chem. Res. Toxicol.

Guiffrida, F. (2004). Formation and hydrolysis of triacylglycerol and sterol epoxydes: Role of unsaturated triacylglycerol peroxyl radicals. *Free Radical Biology & Medicine*, *37*(1), 104–114. doi:10.1016/j.freeradbiomed.2004.04.004 PMID:15183198

Guillen, M. D. (2008). Toxic oxygenated α,β-unsaturated aldehydes and their study in foods. *A Review. Crit. Rev. Food Sci. Nutr.*, *48*, 118–136.

Ignatov, V.I. (2010). Market requirements for special fats and margarine products for industrial use. *Oil and Fat Industry, 1*, 4.

Lvovich, N.A. (2011). The effect of hydrocolloids on the physicochemical properties of Chak-Chak and deep frying fats during frying. *Confectionery Production, 6*, 23-26.

Ma'rques-Ruiz, G. (1996). Quantitation of oxidized triglyceride monomers and dimers as a useful measurement for early and advanced stages of oxidation. *Grasas y Aceites*, *47*, 48–53.

Makarova, A.N. (2011). A study of the effect on the body of snack and butter flour confectionery products during their long-term consumption by the clinical analysis of blood. *Technology and Commodity Science of Innovative Food Products, 3*(8), 67-74.

Markaverich, B. M. (2005). Leukotoxins diols from ground corncob bedding disrupt estrus cyclicity in rats and stimulate MCF-7 breast cancer cell proliferation. *Environmental Health Perspectives*, *113*, 1698–1704.

Marmesat, S. (2007). Used frying fats and oils: Comparison of rapid tests based on chemical and physical oil properties. *International Journal of Food Science & Technology*, *42*, 601–608.

Marmesat, S. (2008). Quantitative determination of epoxy acids, keto acids and hydroxy acids formed in fats and oils at frying temperatures. *Journal of Chromatography. A*, *1211*, 129–134.

Molchanova, N.A. (1980). On the issue of rationing the content of oxidation products in dietary fats. *Commodity Science and light Industry, 7,* 29-33.

Moll, W. F. (1991). Natural Bleaching Clays Time Has Come… Int. News Fats, Oils and Relat. Mater., 2(4).

Moreira, R. G., Castell-Perez, M. E., & Barrufet, M. A. (1999). *Deep fat frying: Fundamentals and applications*. Aspen Publ.

Nosova, A. S. (2013). *Improving the quality of fat for the production of deep frying products. Extended abstract of Cand.Sci. (Eng.)*. Dissertation. Orel.

Nzai, J. M., & Proctor, A. (1998). Determination of Phospholipids in Vegetable Oil by Fourier Transform Infrared Spectroscopy. *Journal of the American Oil Chemists' Society*, *75*(10), 1281–1289.

Okunevich, I. V. (2004). Antioxidants: The effectiveness of natural and synthetic compounds in the complex therapy of cardiovascular diseases. *Reviews on the Clinical, Pharmacological and Medical Therapy*, *3*(3), 2–17.

Perkel, R.L, (2013). The mechanism of formation and dynamics of the accumulation of free fatty acids during frying potatoes. *Problems of Economics and Management in Trade and Industry, 4,* 69-74.

Perkel, R.L. (2014). Safety criteria and methods for evaluating deep-fried fast food products. *Problems of Economics and Management in Trade and Industry, 4*(8), 90-101.

Perkel, R.L. (2019). Rationing and experimental determination of safety indicators for deep-fried yeast dough products. *Technique and Technology of Food Production.*

Porter, N. A. (1995). Mechanisms of free radical oxidation of unsaturated lipids. *Lipids*, *30*(4), 277–290.

Pravisani, C. I., & Calvelo, A. (1986). Minimum cooking time for potato strip frying. *Journal of Food Science*, *51*, 614–617.

Primo-Martin, C. (2011). Deep-fat fried battered snacks prepared using super heated steam (SHS): Crispness and low oil content. *Food Research International*, *44*, 442–448.

Recommendations (2014). "Methodological recommendations for determining the quality of deep frying fat using indicator strips 3M LRSM (indicator of low concentrations of free fatty acids)" No. 17 FTs / 4097 dated 12/31/2004.

Rogozin, I.P. (2018a). Kinetics of changes in fat safety indicators during deep frying of semi-finished dough products. *Actual Biotechnology, 3*(26), 466-470.

Rogozin, I. P. (2018b). Safety assessment of deep frying fat by the content of oxidation products insoluble in petroleum ether. *Science Week SPbPU: Proceedings of a scientific conference with international participation, 2018*, 163–165.

Rogozin, I.P. (2019). Input control of resistance to oxidation of deep frying fats in the production of fast food products from yeast dough. *Bulletin of the Krasnoyarsk State Agrarian University, 4*(145), 138-145.

SanPin 2.3.6.1079-01 "Hygiene and Epidemiology Criteria for Public Catering Establishments and for the Preparation and Circulation By Them of Food Products and Raw Foodstuffs". Moscow, 2001.

Simakova, I. V. (2009). The effect of deep frying fats on the blood composition of experimental animals. *Collection of scientific papers. St. Petersburg, 2009*, 42–47.

Stopsky, V.S. (1983). *Method for the quantitative determination of epoxy groups in fats*. USSR author's certificate No. 1040914, IPC7 G 01 N 33/02, G01N 31/02. Applicant: Scientific Production Association "Maslozhirprom".

Tavassoli-Kafrani, M. H. (2017). The use of ATR-FTIR Spectroscopy to Measure Changes in the Oxirane Content and Iodine Value of Vegetable Oils during Epoxidation. *European Journal of Lipid Science and Technology, 119*(7), 1600354.

Technical Regulations of the Customs Union TR CU 021/2011 On Food Safety.

The Manual. (1967). *Manual to research methods, technological control and accounting of production in the oil and fat industry* (vol. 1). Academic Press.

Tseng, Y. C., Moreira, R., & Sun, X. (1996). Total frying-use time effects on soybean-oil deterioration and on tortilla chip quality. *International Journal of Food Science & Technology*, *31*, 287–294.

Varela, P. (2011). Hydrocolloids in Fried Foods. A Review. *Food Hydrocolloids*, *25*, 1801–1812.

Vaskina, V. A. (2014). The use of hydrocolloids as surface anti-fat barriers. *Confectionery and Bakery Production*, (1-2), 18–21.

Velasco, J. (2004). Formation and evolution of monoepoxy fatty acids in thermooxidized olive and sunflower oils and quantitation in used frying oils from restaurants and fried-food outlets. *Journal of Agricultural and Food Chemistry*, *52*, 4438–4443.

Wilson, R. (2002). Dietary epoxy fatty acids are absorbed in healthy women. *European Journal of Clinical Investigation*, *32*(2), 79–83.

Yates, R. A. (1992). Adsorptive Capacity of Active Filter Aids for Used Cooking Oil. *Journal of the American Oil Chemists' Society*, *69*(9), 894–897.

Chapter 6
Innovative Deep-Fried Fast Food Production Control Scheme

ABSTRACT

In this chapter, the authors analyzed the process of fast-food production using the example of enterprises that use deep frying and produce products with a large proportion of fat. The safety indicators of fats and regulatory documents for the control of the safety of fast-food production are analyzed. For the first time, a new scheme for monitoring the safety of fast-food products has been proposed, which provides for monitoring the indicators of finished products. The standards of safety indicators at critical control points of the production process are proposed.

INTRODUCTION OF THE CHAPTER

The materials in this chapter are based entirely on the authors' own research.

The purpose of this research was to develop an innovative scheme for the production control of fast food products.

The expediency of conducting research on this chapter is dictated by the lack of proven effective production control schemes and safety management systems for fast food products. Fast food deep-frying companies control only the safety of the fat component at the deep-frying stage. Until now, safety

DOI: 10.4018/978-1-7998-8197-1.ch006

control of the finished product itself and its fatty component at the stages of sale and circulation is not provided for by regulatory documents.

MAIN FOCUS OF THE CHAPTER

6.1. Optimization of Fatty Acid Composition of Deep Frying Fats

To ensure biological usefulness, modern deep frying fats must meet the requirements of Table 1.

Table 1. The composition of fat products for nutrition of a healthy person in accordance with the recommendations of Rospotrebnadzor 2.3.6.2432-08

The degree of unsaturation of fatty acids	The ratio of fatty acids, % of the total energy of daily diet	The ratio of fatty acids, % of the total fat
Saturated	Not more than 10	38.5 - 33.3
Monounsaturated	10	38.5 - 33.3
Polyunsaturated	From 6 to 10	23.0 - 33.3
Including: polyunsaturated acids ω6 polyunsaturated acids ω3	From 5.6 to 8.3 From 0.4to 1.7	21.6 - 27.8 1.4 - 5.5

In order to increase the oxidative stability of deep frying fats, it is advisable to choose a variant with a minimum content of polyunsaturated fatty acids (20-25%) and saturated fatty acids (not more than 33-35%). This is especially true for mixtures containing natural palm oil. In these mixtures, the required melting temperature of deep frying fat of not more than 36 °C is achieved when the palm oil content in the mixture is not more than 60%.

In addition, in accordance with the requirements of the Technical Regulations of the Customs Union TR CU 024/2011, from 01.01.2018, edible fats should contain no more than 2% of trans-isomers of unsaturated fatty acids of the total fatty acids. In accordance with the requirements of this regulatory document, hydrogenated vegetable oils cannot be used in the production of edible fats.

Table 2. The fatty acid composition of natural vegetable oils used in the Russian Federation for the production of deep frying fats (Personal research data)

Name and notation of fatty acid	The fatty acid composition of deep frying fats, mass fraction, %			
	Sunflower oil	High oleic sunfloweroil	Palmoil	Palm olein
Lauricacid C12:0	-	-	0,2	0,2
Myristicacid C14:0	0.1	-	1.0	1.0
Palmiticacid C16:0	6.4	4.3	45.1	41.0
Palmitoleicacid C16:1	-	0.2	0.1	-
Stearicacid C18:0	4.1	2.8	4.1	4.0
Oleic acid C18:1	22.8	82.0	39.0	41.7
Linoleicacid C18:2	65.5	9.4	9.8	10.9
Linolenoicacid C18:3	0.1	-	0.3	0.2
Arachicacid C20:0	0.2	0.2	0.3	0.3
Eicosenicacid C20:1	0.1	0.3	-	0.1
Behenicacid C22:0	0.7	0.1	0.1	0.1
Sum of saturated fatty acids (SFA) Sum of monounsaturated fatty acids Sum of polyunsaturated fatty acids	11.4 23.0 65.6	7.4 83.2 9.4	50.8 39.1 10.1	46.6 42.3 11.1

Table 2 shows the fatty acid composition of natural vegetable oils used in the Russian Federation for the production of deep frying fats.

Since 2018, new stringent requirements for the content of carcinogenic esters of glycidol (1,2-epoxypropanol) and 3-MCPD (3-monochloropropanediol) esters in refined vegetable oils have been set in the countries of the European Union. In accordance with the regulations of the European Union, the established standards are not more than 1 mg/kg of these compounds by weight of vegetable oils. Accordingly, similar standards are set for the content of glycidol and 3-monochloropropanediol esters in edible fats derived from vegetable oils.

In August 2019, the Commission of the Customs Union in the Russian Federation invited the Government of the Russian Federation to introduce similar requirements into the Technical Regulations of the Customs Union for the production of refined deodorized vegetable oils and edible fats. To date, this issue is under consideration, but it is expected that similar standards will be adopted in 2020.

Currently, domestic-made refined deodorized sunflower oil contains about 1-2 mg/kg of glycidol and 3-monochloropropanediol esters. Palm oil and its fractions purchased in Malaysia and Indonesia contain from 8 to 12 mg/kg of these compounds.

The Malaysian Union of Palm Oil Producers proposes to supply a certain amount of palm oil with the required content of glycidol and 3-monochloropropanediol esters to the Russian Federation but at higher prices.

"Bernardini" (Italy) has proposed a new technology for processing vegetable oils to produce samples with the required content of glycidol and 3-monochloropropanediol esters.

According to the data given in Section 4.3.3, the content of epoxy compounds, mainly epoxy acids containing an epoxy group in the fatty acid chain, in used deep frying fats and oils reaches at least 65 meq/kg, that is, at least 2% and even up to 3% by weight of fat (up to 30 g/kg) (Velasco, J., 2004).

Recent studies have proven that these epoxy acids do have carcinogenic properties (Greene J.F., 2000; Wilson R., 2002; Markaverich B.M., 2005), while their content in spent deep fat is 2-3 thousand times higher than the normalized content of glycidol esters containing an epoxy group in the skeleton of propanediol.

Chapter 6 of this work presents the results of biological tests of deep frying fats in the nutrition of laboratory animals - white rats. Vegetable fats and oils that did not undergo special purification to remove glycidol and 3-monochloropropanediol esters from them were used for the production of experimental samples of deep frying fats.Nevertheless, when animals were fed with initial fat mixtures and up to the use of experimental samples of oxidized deep-fat fats containing up to 19 g/kg of epoxy compounds in fatty acid chains, there were no changes in blood parameters of experimental animals characteristic of samples with carcinogenic properties. Thus, this issue needs further study and scientific justification.

Analysis of Tables 6.1 and 6.2 show that mixtures of ordinary sunflower oil with palm oil and palm olein, mixtures of high oleic sunflower oil with palm oil and palm olein, as well as high oleic sunflower oil and its mixtures with ordinary sunflower oil are suited for the production of deep frying fats that meet the above requirements.

To optimize the fatty acid composition of the designed deep frying fats, the mass fraction of SFA and PUFA in two-component mixtures of these oils and fats was calculated at a content of each component from 0 to 100%.

The content of any fatty acid in the mixture can be calculated by the following formula:

$FA_m = FA_1M_1/100 + FA_2M_2/100 + \ldots FA_nM_n/100$,

where FA_m, FA_1, FA_2, ... FA_n is the concentration of this fatty acid in the mixture and in each component of the mixture, respectively, %;

M_1, M_2, ..., M_n is the mass fraction of this fatty acid in each component of the mixture, respectively, %. Sum of M_1, M_2, ..., M_n is 100%.

For two-component mixtures, this formula takes the following form:

$FA_m = FA_1M_1/100 + FA_2M_2/100$,

where the notation is the same as in the previous formula.

After some manipulations we get:

$FA_m = FA_1(100 - M_2)/100 + FA_2M_2/100 = FA_1 + (FA_2 - FA_1)M_2/100$.

Using this formula, we compiled equations to calculate the mass fraction of saturated fatty acids (SFA, %) and unsaturated fatty acids (PUFA, %) in mixtures of ordinary sunflower oil with palm oil, depending on the mass fraction of palm oil in the mixture (M_2, %) (Table 3):

$SFA = 11.4 + (50.8 - 11.4)M_2/100 = 11.4 + 0.394M_2\%$;

$PUFA = 65.6 - (65.6 - 10.1)M_2/100 = 65.6 - 0.555M_{2\%}$.

If SFA = 35% and PUFA = 20-25%, a joint solution of these equations is impossible.

Table 3. Mass fraction of saturated fatty acids and unsaturated fatty acids in mixtures of ordinary sunflower oil with palm oil (Personal research data)

Mass fraction of palm oil in mixture, %	Mass fraction of saturated fatty acids, % of fatty acids sum	
	saturatedfattyacids	polyunsaturatedfattyacids
0	11.4	65.6
10	15.3	60.1
20	19.3	54.5
30	23.2	49.0
40	27.2	43.4
50	35.0	37.9
60	34.5	32.3
70	39.0	26.8
80	42.9	21.2
90	46.9	15.7
100	50.8	10.1

A mixture of ordinary sunflower oil with palm oil, containing 40-60% of palm oil partially satisfies the specified requirements. Such mixtures contain from 27.2 to 34.5% of SFA and from 43.4 to 32.3% of PUFA. To achieve the required thermal stability, these mixtures must be enriched with effective antioxidant complexes.

Table 4. Mass fraction of saturated fatty acids and unsaturated fatty acids in mixtures of ordinary sunflower oil with palm olein (Personal research data)

Mass fraction of palm olein in mixture, %	Mass fraction of saturated fatty acids, % offattyacidssum	
	saturatedfattyacids	polyunsaturatedfattyacids
0	11.4	65.6
10	14.9	60.2
20	18.4	54.7
30	22.0	49.3
40	25.5	43.8
50	29.0	38.4
60	32.5	32.9
70	36.0	27.5
80	39.6	22.0
90	43.1	16.6
100	46.6	11.1

The characteristics of mixtures of ordinary sunflower oil with palm olein are presented in table 4.

Equations for calculating the mass fraction of SFA and PUFA in mixtures:

$SFA = 11.4 + (46.6 - 11.4)\ M_2/100 = 11.4 + 0.352M_{2\%}$;

$PUFA = 65.6 - (65.6 - 11.1)\ M_2/100 = 65.6 - 0.545M_{2\%}$,

whereM_2 is the mass fraction of palm olein, %.

A mixture of ordinary sunflower oil with palm olein, containing 40-60% of palm olein partially satisfies the specified requirements. Such mixtures contain 25.5 - 32.5% of SFA and 43.8 - 32.9% of PUFA and also require additional stabilization with antioxidant additives.

The characteristics of mixtures of high oleic sunflower oil with palm oil are given in Table 5.

Equations for calculating the mass fraction of SFA and PUFA in mixture:

$SFA = 7.4 + (50.8 - 7.4)\ M_2/100 = 7.4 + 0.434M_{2\%}$;

Table 5. Mass fraction of saturated fatty acids and unsaturated fatty acids in mixtures of high oleic sunflower oil with palm oil (Personal research data)

Mass fraction of palm oil in mixture, %	Mass fraction of saturated fatty acids, % offattyacidssum	
	saturatedfattyacids	polyunsaturatedfattyacids
0	7.4	9.4
10	11.7	9.4
20	16.1	9.5
30	20.4	9.6
40	24.8	9.7
50	29.1	9.8
60	33.4	9.8
70	37.8	9.9
80	42.1	10.0
90	46.5	10.0
100	50.8	10.1

$PUFA = 9.4 + (10.1 – 9.4) M_2/100 = 9.4 + 0.007M_{2\%}$.

Mixtures containing less than 60% of palm oil satisfy the above requirements for the content of saturated fatty acids.

Equations for calculating the mass fraction of SFA and PUFA in mixtures of high oleic sunflower oil with palm oil:

$SFA = 7.4 + (46.6 – 7.4) M_2/100 = 7.4 + 0.392M_{2\%}$;

$PUFA = 9.4 + (11.1 – 9.4) M_2/100 = 9.4 + 0.017M_{2\%}$.

Table 6. Massfraction of saturated fatty acids and unsaturated fatty acids in mixtures of high oleic sunflower oil with palm olein (Personal research data)

Mass fraction of palm olein in mixture, %	Mass fraction of saturated fatty acids, % offattyacidssum	
	saturatedfattyacids	polyunsaturatedfattyacids
0	7.4	9.4
10	11.3	9.6
20	15.2	9.7
30	19.2	9.9
40	23.1	10.1
50	27.0	10.3
60	30.9	10.4
70	34.8	10.6
80	38.8	10.8
90	42.7	10.9
100	46.6	11.1

The characteristics of mixtures of high oleic sunflower oil with palm olein are given in Table 6.

Mixtures containing less than 70% of palm olein satisfy the above requirements for the content of SFA. A mixture containing 70% of palm olein and 30% of high oleic sunflower oil contains 34.8% of SFA and 10.6% of PUFA. The mixture has a fairly high thermal stability.

Table 7. Massfraction of saturated fatty acids and unsaturated fatty acids in mixtures of high oleic sunflower oil with ordinary sunflower oil (Personal research data)

Mass fraction of ordinary sunflower oil in mixture, %	Mass fraction of saturated fatty acids, % offattyacidssum	
	saturatedfattyacids	polyunsaturatedfattyacids
0	7.4	9.4
10	7.8	15.0
20	8.2	20.7
30	8.6	26.3
40	9.0	31.9
50	9.4	37.5
60	9.8	43.1
70	10.2	48.8
80	10.6	54.4
90	11.0	60.0
100	11.4	65.6

The characteristics of mixtures of high oleic sunflower oil with ordinary sunflower oil are given in Table 7.

Equations for calculating the mass fraction of SFA and PUFA in mixtures:

$SFA= 7.4 + (11.4 – 7.4) M_2/100 = 7.4 + 0.04 M_{2\%}$;

$PUFA = 9.4 + (65.6 - 9.4) M_2/100 = 9.4 + 0.562 M_{2\%}$.

Any of the above mixtures meets the above requirements for the content of SFAs, requirements for the content of PUFAs - mixtures containing not more than 25% of ordinary sunflower oil.

Three-component mixtures containing 60% of palm oil (or palm olein), 20% of ordinary and 20% of high oleic sunflower oil fully meet the above requirements. These mixtures contain from 34.2 to 31.7% of SFA and from 21.1 to 21.7% of PUFA.

The specific ratios of components in mixtures that meet the specified requirements for fatty acid composition are determined by current prices of individual oils.

The thermal stability of mixtures containing a higher level of PUFA can be improved by using approved food additives that inhibit oxidative processes in deep frying fat - antioxidants, synergists, antifoam agents, and complexing agents. In this case, preference should be given to the use of natural antioxidant complexes containing tocopherols, tocotrienols, ascorbic acid and ascorbyl palmitate, citric acid, and bioflavonoids.

Thus, mixtures of natural fats and oils containing not more than 2% of trans-isomers of fatty acids, not more than 35% of SFA and 20-25% of linoleic acid that meet the FAO / WHO recommendations for a healthy diet are recommended for the production of deep frying fats.In particular, mixtures of sunflower oil and high oleic sunflower oil with palm oil and palm olein are suitable for this purpose. With a higher content of linoleic acid, additional stabilization of deep frying fat by natural antioxidant complexes is recommended.

In Finland, Lithuania and other European countries, low erucic rapeseed oil is widely used for the production of deep frying fats. This oil contains 15-16% of saturated fatty acids, 53-54% of monounsaturated fatty acids, 31-32% of polyunsaturated fatty acids, including 5-7% of linolenic acid. Mixtures of 75% low erucic rapeseed oil with 25% of ordinary sunflower oil are successfully used in these countries as deep frying fats with the addition of antioxidant additives. Despite the presence of 3-5% of linolenic acid in the mixtures, the thermal stability of the mixtures is quite high and is about 7 hours.

To slow down the development of oxidation processes in fat-containing foods or edible fats, it is necessary to exclude the influence of factors contributing to the development of a radical chain oxidation process in food production.

For frying foods, and especially for repeated deep-frying, it is necessary to use thermostable cooking and deep frying fats with the possibly low content of polyunsaturated fatty acids. Fats and oils should be stored before use in a sealed container, without access of oxygen and light.

When frying foods in deep frying fat, foaming of fat should be avoided, since it significantly increases the contact surface of hot fat with oxygen. For these purposes, TR CU 029/2012 mandates the use of special food additives - antifoam agents, for example, polydimethylsiloxane (E 900), which in concentrations of 1...2 mg/kg of fat forms a monomolecular film on its surface that prevents the contact of fat with oxygen.

The chain reaction rate can be reduced by converting the active radicals R•, ROO•, RO•, •OH to an inactive molecular state. This is achieved by using approved food additives - inhibitors, which react with active free radicals, converting radicals to inactive molecular products and transferring the function of the radical to the inhibitor (reactions 7, 8).

6.2. Entrance Control of Safety Indicators of Deep Frying Fats by The Ransimetry Method. The Relationship Between Resistance to Oxidation and Safety Indicators of Deep Frying Fats in The Technological Process

The purpose of this section was to evaluate the thermal stability of some industrially produced deep frying fats of domestic and foreign production used in the production of fast food products from yeast dough by a frying method. Tasks of the work:

1. Investigate the oxidation stability of some industrially produced deep frying fats by a ransimetry method.
2. To establish the relationship between the resistance to oxidation of deep frying fats and the maximum allowable their usage time.

The objects of the study were:

Deep frying fats "Sunny Gold" and "Vegafray 05" produced by LLC "Cargill Russia".

Frying fat "Rainbow" produced by Rucola JSC Eurasian Conformity (Lithuania).

The resistance to oxidation of deep frying fats was evaluated by the ransimetry method (accelerated oxidation method) in accordance with GOST R 51481-99 (ISO 6886-96). The fat was oxidized with atmospheric oxygen at 110 °C in oxygen kinetic mode. The oxidation rate was estimated by the accumulation of volatile low molecular weight fatty acids, which was controlled by the conductometric method taking into account increase in the conductivity of the solution in the measuring cell.The change in the conductivity of the solution in the measuring cell was recorded automatically depending on the duration of oxidation in the form of a conductivity curve (Figure 1).

The duration of the induction period was determined by the maximum on the second derivative of the conductivity curve (Rogozin, 2019).

The thermal stability of deep frying fats in a real technological process was studied during oxidation at frying of dough semi-finished products (cakes, donuts, battered fruits and vegetables) at 180 °C. The oxidation of fat was carried out in a diffusion mode.

Deep frying products were fried for 8 hours with sampling of deep-fat fat every 1-2 hours. The content of conjugated dienes, the content of compounds insoluble in petroleum ether (SPSE), and the content of epoxides by the above methods were determined in the studied deep-fat fat samples.

Frying products was carried out during 8 hours with deep frying fat sampling every 1-2 hours. Content of interfaced dienes, the content of compounds

Figure 1. The conductivity curve of the solution in the measuring cell during accelerated oxidation of the "Rainbow" deep frying fat and its second derivative. Fat oxidation stability is 6.1 hours (Personal research data)

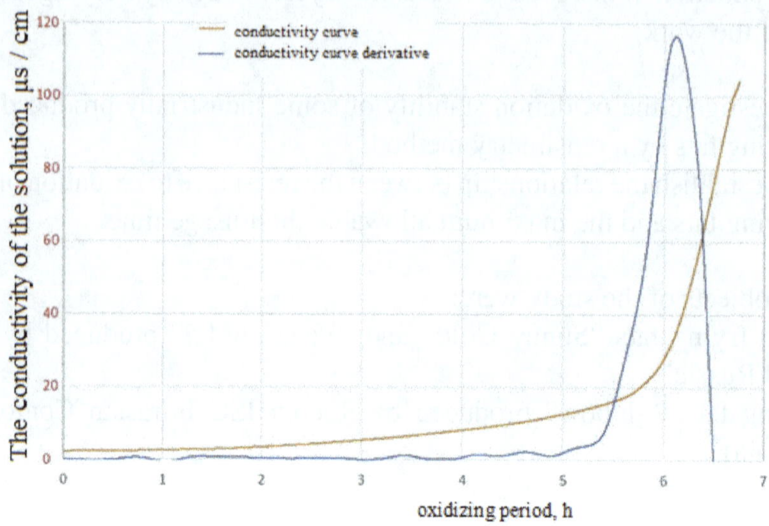

insoluble in petroleum ether, and the content of epoxides were determined in studied samples of deep frying fats.

In accordance with working SanPin 2.3.6.1079-01, the maximum permissible period of use of deep frying fat ended when at least 1.0% of compounds insoluble in petroleum ether and 65 meq/kg of epoxides accumulated in the fat.

Table 8 shows the results of determining the resistance to oxidation of fat for the initial samples of deep frying fats “Sunny Gold”, “Vegafray 05” and “Rainbow”, as well as the resistance to oxidation of these fats samples after their prolonged use in the technological process.

This table shows the results of determining the thermal stability of fats by the accumulation of products of thermo-oxidative destruction of deep frying fat. For example, the thermal stability of the “Sunny Gold” deep frying fat sample was about 5.5 hours, and the thermal stability of “Rainbow” deep frying fat was 7 hours.

Table 8. Oxidation resistance and real thermal stability of deep frying fats when deep frying yeast dough products (Personal research data)

Characteristic of deep frying fats	Oxidation resistance of deep frying fats according to the ransimetry data, h	Real thermal stability of deep frying fats when deep frying yeast dough products		
		Mass fraction of conjugated dienes, meq/kg	Mass fraction of epoxides, meq/kg of fat	Mass fraction of compounds insoluble in petroleum ether in fat, %
“Sunny Gold” fat, initial	4.4	30.4	9.8	0.17
“Sunny Gold” fa, after use for 6 hours	-	59.3	76	1.14
“Rainbow” fat, initial	6.1	21.4	6	0.37
“Rainbow” fat, after use for 7 hours	-	57.1	65	1.0

In a similar experiment, the resistance to oxidation of “Vegafray 05” deep frying fat was 11.1 hours; thermal stability was about 13 hours [Rogozin, 2019].

Evaluation of the results allows one to conclude that there is correlation dependence between the oxidation stability of deep frying fat, determined by the ransimetry method, and thermal stability of fat by the accumulation of oxidation products in it. This correlation is expressed by equation:

$T_{d=} 1.2T_r$,

where T_d is a thermal stability of deep frying fat, h;

T_r is an oxidation stability of deep frying fat according to the ransimetry data, h.

The coefficient 1.2 reflects the various conditions of fat oxidation by atmospheric oxygen and the control of the oxidation process by the accumulation of various final products. When investigated by the ransimetry method, fat oxidation occurs at 110 °C in the kinetic mode, and in a real technological process - at 180 °C in the oxygen diffusion mode.

When calculating according to the results of ransimetry method, the permissible time for using "Sunny Gold" deep frying fat in the technological process is T_d = 1.2××4.4 = 5.3 hours, which matches the data given in the Table 6.8.

Preliminary testing of each new batch of deep frying fat using the ransimetry method allows one to calculate the allowable use duration of this fat in the process without additional physical and chemical analyzes.

To verify the results, the monitoring of the content of oxidation products when replacing used deep frying fat with fresh one can be carried out periodically in accordance with the approved control scheme, especially at the use of a new batch of deep frying fat.

Compulsory input control of the thermal stability of each batch of deep frying fats by accelerated oxidation method will allow one to control the permissible use period of these fats in the production process and replace the used deep frying fat with fresh one in a predetermined time without any additional analyzes. Deep frying fats with an oxidation stability of at least 4 hours are recommended for use in an industrial process.

6.3 Safety Indicators of Deep Frying Fats and Production Control at Enterprises Producing Fast Food Products with A High Content of Fat Component

The HACCP system focuses on the prevention of risks and ensuring the safety of food products at all stages of the production process - from the production of raw materials to the release of finished products. In this system, possible risks are predicted in advance, and the places where such risks may arise are identified as critical control points.

Russian enterprises that produce deep-fried fast-food products are distinguished by the fact that they control only the safety of deep frying fat in the process. Until now, the safety control of the finished product and the fat component of this product at the final stage of the technical process and the circulation stage are not provided for in regulatory documents.

The production process of enterprises producing fast food products proceeds either in an accelerated mode (fast food network enterprises) or in a process flow mode (industrial enterprises), which makes it difficult to monitor the safety indicators of products.

As a rule, the technological process of such enterprises, is standardized, this allows one to reduce the time for personnel training and to use lower-skilled workers. The range of products is small.

In order for the consumer to receive a finished product that meets the safety requirements, the raw material must pass safety control at all stages of the technological process - from the input of raw materials to their processing and sale at the manufacturing facility (catering) or in the distribution network.

Currently, the input safety control of raw materials is carried out only by analysis of supporting documents and visually - by organoleptic indicators. This is not always enough to ensure product safety.

Specific studies may be required to verify the effectiveness of the supplier's HACCP system. It is this stage that is basic in identifying and analyzing potential risks.

Potential risks are monitored at critical control points in accordance with the HACCP plan developed at the enterprise.

The most labile component of fast food is fat. As a rule, this is a special-purpose fat for deep frying or a functional and technological ingredient in fast food products.

The HACCP plan for the production of fast food products with a large proportion of the fat component must necessarily take into account the risk of lipid oxidation at the stage of production and circulation; therefore, monitoring the compliance with the safety standards of deep frying fats and finished products is an important task.

In accordance with the regulatory documents of the Russian Federation (TR CU 021/2011 and TR CU 024/2011), the safety of the fat component of these products is determined by the content of peroxides (not more than 10 meq/kg) and free fatty acids (FFA) - not more than 0.6 mg KOH / g.

Table 9. Description of deep frying fat (Personal research data)

Deep frying fat			
Check sheet	**Components/ indicators**	**Norm**	**Resource**
1	**2**	**3**	**4**
1. Name of raw material	Deep frying fat		
2. Raw materialcomponents	Fat (99.8%) Water (0.2%)		
3. Safetyindicators	Physical and chemical indicators, max: *1.Trans-isomers of fatty acids, %* 2.Acid number, mgKOH/g 3.Peroxide number, mEq/kg	2 0.6 10	TR CU 024/2011 Certificate of safety
	Mass content of compounds insoluble in petroleum ether, %, max Mass content of epoxides, meq/kg, max	0.15 10	Notapplicable
	Microbiological indicators: pathogenic micro-organisms, including Salmonella in 25.0 g QMAFAnM, CFU/g (cm^3), max Coliform bacterias in 1 g of product, In retail packaging S.aureus in 0.1 g (cm^3) yeast, CFU/g (cm^3) mold, CFU/g (cm^3)	Not allowed Not allowed 0.001 Not allowed 1×10^3 1×10^2	TR CU 021 / 2011 TR CU 024/2011 SanPin 2.3.2.1078-01 Certificate of safety
	Toxic elements: PB, mg/kg, max As, mg/kg, max Cd, mg/kg, max Hg, mg/kg, max	0.1 0.1 0.05 0.05	TR CU 021/ 2011 Certificate of safety
	Mycotoxins: aflatoxin M_1, mg/kg, max	0.005	TRCU 021 / 2011
	Pesticides: HCCH (α-, β-, γ -isomers), mg/kg, max DDT and its metabolits, mg/kg, max	0.05 0.1	TRCU 021/ 2011
	Dioxines, max	0.00000075	TRCU 021/2011
	Radionuclide content, Bq/kg, max: -Cz-137; -Sr-90	60 80	TRCU 021 /2011
4. General properties of raw material	Organoleptic indicators: appearance and consistency	Conform to the composition of the used fats and the standard of the manufacturing facility	
5. Packaging	Retailpackaging	Laminated aluminum foil or its substitutes, glass, plastic	TRCU 005/2011

Table 9 shows an example of a description of deep frying fat as the main raw material for the production of fast food products, at the stage of its input control.

As can be seen from the data in Table 6.9, all monitored safety indicators must comply with the requirements of the Technical Regulations of the Customs Union.

However, the safety indicators of the fat component regulated by TR CU 021/2011 and TR CU 024/2011 do not fully reflect the safety requirements for the fat component, since there is no most important indicator of fat safety - the content of secondary oxidation products insoluble in petroleum ether. Their maximum regulated concentration when replacing the used deep frying fats with fresh one in accordance with SanPin 2.3.6.1079-01 is 1%.

Based on the above studies, an innovative control scheme for the fast food production process has been developed, providing for a real definition of the toxic products of deep frying fat oxidation in the finished product.

It is proposed to introduce the determination of the content of secondary oxidation products insoluble in petroleum ether and epoxides in the initial deep frying fat at the stage of the input control of deep frying fats in the food industry enterprises (Table 6.9).

The mass fraction of secondary oxidation products insoluble in petroleum ether should not be higher than 0.15%, the mass fraction of epoxides should not be higher than 10 meq/kg.

In addition, at the stage of input control, it is necessary to determine the real resistance of the used deep frying fat to oxidation by the ransimetry method. It is proposed to use deep frying fats with an oxidation stability of at least 4 hours [Rogozin I.P., 2019]. In the future, it is necessary to provide an indicator of the stability of frying fat to oxidation at a temperature of 110 °C in hours in the certificate of the supplier of deep frying fat, as is provided abroad for other culinary, deep frying and confectionery fats.

Entrance control of the stability of deep frying fat to oxidation by the ransimetry method must be carried out for each new batch of deep frying fat.

Based on the results of ransimetry, the allowable time for deep frying fat using in the technological process is calculated, after which the deep frying fat indicators will not be higher than those stipulated by the regulatory documents of the enterprise; the mass fraction of secondary oxidation products insoluble in petroleum ether in deep frying fat is not more than 1.0%, the epoxy content is not more than 65 meq kg.

The allowable time for deep frying fat using in the technological process in hours is calculated according to the above formula:

$T_{d=} 1{,}2T_r$.

After the raw materials inputted at the enterprise, it is necessary to carry out its storage under controlled conditions. Correctly set and controlled temperature conditions in the refrigerator allow preserving the safety of raw materials.

Further, the raw material goes to production sites and / or production lines where heat treatment takes place.

In the process, a number of factors can affect product safety.

First of all, employees of the enterprise should know:

a list of the hazards that are controlled at workplaces;

values of permissible critical limits of indicators at critical control points (CCP);

applicable monitoring procedures;

records of indicators that need to be saved;

corrections to be taken in cases of deviation of the technological process from the norm.

In addition, it is necessary to monitor compliance with personal hygiene standards.

Secondly, in the production of fast food products, special attention should be paid to observing the parameters of the deep-frying technological process, depending on the type of product being fried.

It is necessary to use thermostable deep frying fat, for which the time of use in the technological process is established by the ransimetry method. Quality control of finished products is carried out using the organoleptic method according to the rating scale adopted at the enterprise.

When setting up the technological process, at the replacing used deep frying fat with the fresh one or with deep frying fat with other characteristics, the its stability to oxidation is determined by the ransimetry method and the time of fat using in the technological process is recalculated.

Verification of the correct use of deep frying fat is carried out after replacing it with fresh fat by studying the indicators of the content of oxidation products in the used fat. The mass fraction of secondary oxidation products insoluble in petroleum ether in the used fat when its replacing should be no more than 1.0%, the content of epoxides should be no more than 65 meq/kg. At significant deviations of the regulated and actual indicators, they should be taken measures to clarify the estimated time factor for the fat use. It can be fixed, for example, from a value of 1.2 to a value of 1.1.

Considering that the determination of secondary oxidation products insoluble in petroleum ether is carried out by a rather laborious and time-consuming method, it is recommended to choose the method of epoxides determination, since both of these indicators are correlated. The methodology for epoxides determining requires significantly less time and is carried out using relatively simple laboratory equipment.

Verification of the proposed control method is carried out periodically in accordance with the standard of the enterprise, for example, once every two weeks. It is necessary at replacing fat with a new batch of deep frying fat, with new raw materials or at production of a new type of product.

The new control method does not require any analytical definitions during the technological process. At the moment of replacing the used deep frying fat with fresh one, the content of oxidation products in the used deep frying fat is guaranteed not higher than the established norms, and in the previous period of use it should be less than these norms.

The safety indicators of the finished product must be monitored in the following order.

First, the fat content in the finished product is determined by the extraction-weight method according to GOST R 54607.8-2016 or by a specially developed method of solid-phase extraction (Perkel R.L., 2019).

To isolate fat, 10-30 g of an average sample of the finished product are carefully ground using a blender. The crushed product is mixed with a prepared desiccant (calcined crushed powder of sodium carbonate, sodium sulfate or calcium chloride) in a ratio of 4: 1 to 2: 1 for the most complete moisture absorption.

The resulting dried mass is transferred to a polyethylene cartridge for solid-phase extraction, on the bottom of which a circle of filter paper is placed. The mass was carefully sealed with a piston to ensure uniform filling of the cartridge. Chloroform (trichloromethane) is added in small portions from above in an amount of 10-30 cm^3 to extract fat.

Fat is eluted by passing the solvent through the sample at a rate of about 1 cm^3/min. The end of the fat extraction is noted by the disappearance of the greasy trace of the eluate on filter paper. The eluate is collected in a pre-weighed flask of a rotary evaporator. At the end of the fat elution, chloroform is removed from the extract on a rotary evaporator until the odor disappears. The mass of extracted fat in the flask of the evaporator is determined by the difference.

Studies have shown that the solid-phase extraction method ensures the fat extraction in the amount of 95–97% to the fat content in the test sample. If necessary, the corresponding correction factor can be used to calculate the results of the analysis. The duration of preparation and elution of the fat sample is 60-70 minutes, depending on the mass of the test sample.

The fat content in the product is expressed in mass fractions.

The maximum possible percentage mass fraction of oxidation products in the finished product is calculated by the formula:

$(CIPE)_{max}=1.0G$,

where 1.0 is the maximum possible mass fraction of secondary oxidation products insoluble in petroleum ether in deep frying fat;

G is the mass fraction of fat in the product, %.

As follows from this formula, when the frying fat is changed within the prescribed period, no more than 1% of secondary oxidation products insoluble in petroleum ether accumulates in it. In this case, the control is reduced to determining the fat mass in the product. If the technological process of deep frying products is streamlined, the minimum and maximum proportion of fat in the product is usually known. In this case, it is not necessary to determine the fat mass in the product; the calculated maximum possible mass fraction of secondary oxidation products insoluble in petroleum ether in the product can be determined by formula (21), substituting the maximum possible mass fraction of fat into it.

Finished products are stored in specialized equipment with a controlled temperature regime. Storage temperature and shelf life of products must comply with the enterprise standard.

The finished product is packaged in accordance with the packaging requirements for this type of fast food product, and the packaging must be safe.

When products are released outside the manufacturing facility, fast food packaging is marked or a label must be affixed containing the following information: name of the product, name of the regulatory or technical document, according to which the product was produced, storage conditions and expiration date, weight of the packaging unit.

Figure 2. Schematic diagram and critical points of production control on the example of fast food products (Personal research data)

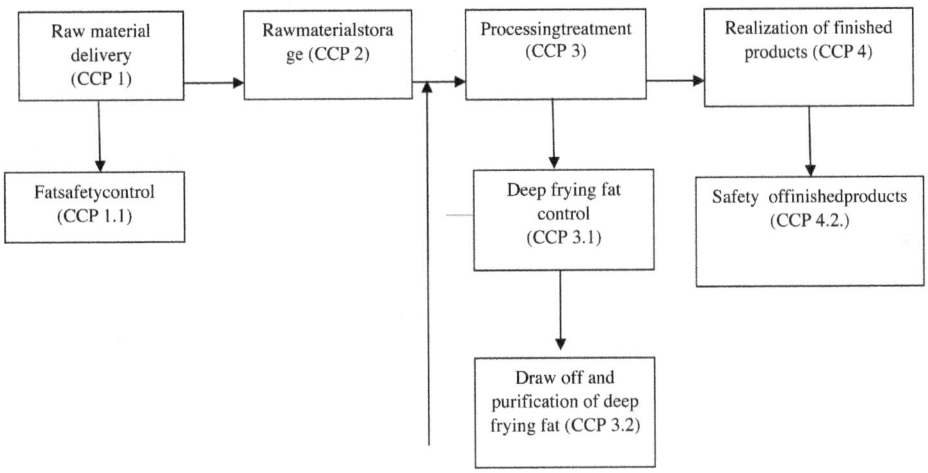

Figure 2 shows the proposed concept of production control and critical control points of the technological process on the example of the production of deep fried fast food products with a large proportion of fat component.

The production control standards for these products are given in Table 10.

It is advisable to filter the used fat daily from burnt and charred particles of the fried product and to replace the used fat with fresh one in a timely manner. Deep frying fat is unsuitable for further use when the estimated time for its use has expired, while the content of compounds insoluble in petroleum ether exceeds 1%.

To reduce the concentration of compounds insoluble in petroleum ether in the used deep frying fat for its further technical use, it is recommended to use adsorption purification of fat, which has not yet been widely recognized in Russia and other countries of the world.

Taking into account the current standards for the content of compounds insoluble in petroleum ether in deep frying fat and an approximate permissible fat absorption level of the finished product of 20%, it is recommended that the mass fraction of compounds insoluble in petroleum ether in the finished product is not more than 0.2% by weight of the product, the mass fraction of epoxides is not more than 13 meq/kg.

If fast food products are not deep fried, but contain a significant amount of fat component, it is necessary to control the safety of the fat component extracted from this product according to the indicators proposed in the table:

Table 10. Norms of production control of the technological process of production of deep fried products (Personal research data)

Critical control points	Control objects	Indicators and norms of control
1	2	3
CCP 1 (Raw material delivery)	Deep frying fat	Safety indicators according to the certificate or according to the declaration of conformance: acid number is not more than 0.6 mgKOH/g, peroxide number is not more than 10 meqO_2/kg, mass fraction of secondary oxidation products insoluble in petroleum ether is not more than 0.15%, mass fraction of epoxides is not more than 10 meq/kg. The content of trans-isomers is not more than 2%. Oxidation stability of deep frying fat according to ransimetry data is at least 4 hours. Organoleptic indicators correspond to the indicators of a fresh product
	Semi-finished product or raw material used for deep-frying of fast food	Safety indicators according to the certificate or according to the declaration of conformance, organoleptic indicators correspond to the indicators of a fresh product
CCP 2 (Rawmaterialstorage)	Deep frying fat	Storage at the temperature of not more than 20°C in a sealed container
	Semi-finished product or raw material used for deep-frying of fast food	Storage under the conditions provided for by the HACCP plan, and according to labeling requirements. Organoleptic indicators correspond to the indicators of a fresh product
CCP 3 (Processcontrol)	Deep-friedfastfoodproduction	Requirements for the temperature and duration of one batch are set by the standard of the enterprise; the actual duration of frying is determined by organoleptic indicators. The time of use of deep frying fat before its replacement with fresh one is calculated according to the data of ransimetry method
CCP 3.1 (Deep frying fat control)	Deep frying fat	Organoleptic indicators correspond to the стандарту standard of the enterprise
CCP 3.2 (Draw off and purification of deep frying fat. Periodic verification of the terms ofdeep frying fat use	Used deep frying fat	During verification, the mass fraction of secondary oxidation products insoluble in petroleum ether is not more than 1%, the mass fraction of epoxides is not more than 65 meq/kg. After adsorption purification, the mass fraction of secondary oxidation products insoluble in petroleum ether is not more than 0.3-0.4%. Organoleptic indicators correspond to the стандарту standard of the enterprise
CCP 4 (Realization of finished products)	Deep fried fast food products	The mass fraction of secondary oxidation products insoluble in petroleum ether is not more than 0.2%, %, the mass fraction of epoxides is not more than 13 meq/kg. The mass fraction of fat is not more than 20%. Organoleptic indicators correspond to the стандарту standard of the enterprise

mass fraction of compounds, insoluble in petroleum ether is not more than 1%, acid number is not more than 2.0 mg KOH/g, peroxide number not more than 10 meq O_2/kg.

CONCLUSION OF THE CHAPTER

When performing research on these sections of the work, the following results were obtained:

1. The main safety criterion for deep-fried fast food products is the minimum content of toxic products of thermo-oxidative degradation of deep frying fats - compounds insoluble in petroleum ether and epoxides. To ensure the safety of this product, it is necessary to control the content of toxic products of deep frying fat oxidation directly in the finished product.
2. The processes of thermal oxidation of deep frying fat at 170-180 ° C differ from the usual process of fats autooxidation at moderate temperatures due to the increased oxidative activity of the LOO• peroxyl radical and LOOH hydroperoxides. As a result of oxidation, deep frying fats accumulate epoxides and dihydroxy compound insoluble in petroleum ether. A correlation between the content of epoxides and compounds insoluble in petroleum ether was established.

Peroxides at temperatures of 170-180 ° C are intermediate products of oxidation and cannot serve to control the degree of oxidation of deep frying fat.

3. Free fatty acids are formed only by hydrolysis of deep frying fat. The accumulation of free fatty acids correlates with the content of polar compounds determined by the electrophysical method. However, neither the determination of acid number using indicator strips, nor the determination of polar products can be used to determine the oxidation degree of deep frying fat.
4. It has been established a correlation between the stability of deep frying fat to oxidation, determined by the ransimetry method, and the permissible duration of deep frying fat use in the technological process. The developed method allows one to anticipate the time of replacing the used fried fat with fresh one. Obtained results allowed offering a fundamentally new scheme of production control of deep fried products.

A method to determine the safety indicators of deep-fried products, based on the control of the mass fraction of fat in these products has been developed.

REFERENCES

Greene, J. F., Newman, J. W., Williamson, K. C., & Hammock, B. D. (2000). Toxicity of epoxy acids and related compounds to cells expressing human soluble epoxide hydrolase. *Chemical Research in Toxicology*, *13*(4), 217–226. doi:10.1021/tx990162c PMID:10775319

Markaverich, B. M., Crowley, J. R., Alejandro, M. A., Shoulars, K., Casajuna, N., Mani, S., Reyna, A., & Sharp, J. (2005). Leukotoxins diols from ground corncob bedding disrupt estrus cyclicity in rats and stimulate MCF-7 breast cancer cell proliferation. *Environmental Health Perspectives*, *113*(12), 1698–1704. doi:10.1289/ehp.8231 PMID:16330350

Perkel, R.L. (2019). Rationing and experimental determination of safety indicators for deep-fried yeast dough products. *Technique and Technology of Food Production*.

Rogozin, I.P. (2019). Input control of resistance to oxidation of deep frying fats in the production of fast food products from yeast dough. *Bulletin of the Krasnoyarsk State Agrarian University, 4*(145), 138-145.

Velasco, J., Marmesat, S., Bordeaux, O., Márquez-Ruiz, G., & Dobarganes, C. (2004). Formation and evolution of monoepoxy fatty acids in thermooxidized olive and sunflower oils and quantitation in used frying oils from restaurants and fried-food outlets. *Journal of Agricultural and Food Chemistry*, *52*(14), 4438–4443. doi:10.1021/jf030753f PMID:15237949

Wilson, R., Fernie, C. E., Scrimgeour, C. M., Lyall, K., Smyth, L., & Riemersma, R. A. (2002). Dietary epoxy fatty acids are absorbed in healthy women. *European Journal of Clinical Investigation*, *32*(2), 79–83. doi:10.1046/j.1365-2362.2002.00951.x PMID:11895453

Chapter 7
Research on the Effects of Fast Food Products on the Body

ABSTRACT

In this chapter, the authors of the study show the results of a series of experiments conducted on more than 200 experimental animals (white rats) with the study of various internal organs, substantiating the negative physiological effect of fat oxidation products. Experiments on animals have proven a close relationship between the content of secondary oxidation products and their effect on the body and the need to regulate and standardize this indicator in finished products at the stages of their circulation and sale to ensure the safety of fast-food products.

INTRODUCTION OF THE CHAPTER

The materials in this chapter are based entirely on the authors' own research.

The purpose of the studies reflected in this chapter was to study the level of toxic effects of oxidation products of fast food products and fats used for frying some of its types, in particular, the concentration of copolymers insoluble in petroleum ether, on the body of animals with prolonged consumption against the background of a balanced diet to establish safety indicators

Special attention was paid to the concentration of compounds insoluble in petroleum ether (CIPE), as well as the content of epoxides and conjugated dienes.

DOI: 10.4018/978-1-7998-8197-1.ch007

The scientific research conducted in this area was dedicated to assessing the safety of deep-frying fats only (mainly sunflower soybean, rapeseed, palm and other oils), subjected to high-temperature processing when frying products in deep-frying fat (Abdel-Aal, M.H., and H.A. Karara. 1986; Andrikopoulos N.K.et al., 2002; Rimac-Brncic, S., et al., 2004; Velasco, J. et al., 2005; Marmesat, S. et al.,2005).

At present, industrial fast food products technology uses a wide range of fats and their mixtures, differing in fatty acid composition, content of approved food additives-antioxidants and quality indicators for which no identical assessment was made in animal experiments, which allows to compare the existing regulated level **CIPE** and their real effect on the body.

In addition, the need for ongoing research is due to the lack of experimental data on the safety and effect on the body made by products with oxidized fat phase containing different concentrations of CIPE, and not deep-frying fats, because a person consumes the product, not the fat, in which this product undergoes thermal processing.

Based on the objectives of this research, the following experimental studies have been conducted:

- Assessment of the safety of industrially produced fast food products with a large proportion of the fat component containing different concentrations of CIPE, both deep-fried and made using other technologies;
- comparison of the experimental data on the intensity of pathological changes in the body, depending on the concentration of CIPE in palm oil, which is widely used as deep-frying fat in food technology;
- a comparative analysis and comparison of the level of toxic effect that the deep-frying fats and products fried in them have on the organism to assess the adequacy of deep frying fats safety indicators monitoring in the technological process;
- Safety assessment of the of the new generation deep-frying fat Vegafray 05, which does not contain trans-isomers, produced by Cargill Russia LLC.

MAIN FOCUS OF THE CHAPTER 7

7.1 Safety Assessment of Industrially Produced Fast Food in Animal Experiments

The research was performed on five groups of rats, including the control group, a part of a full-fledged balanced diet of which was replaced by the most popular fast food products of industrial production (potato chips, crackers, shortbread biscuits, flour product *chak-chak*), the results of the safety indicators study of the fat component are presented in Chapter 5.

The effects on the animals during long-term consumption of experimental diet were studied with patho-morphological, histological and hematological methods. Animal studies were carried out on the basis of a certified vivarium and scientific-technological center of the Veterinary Hospital and the laboratories of the Department of Morphology, Animal Pathology and Biology of the Saratov State University and conducted in accordance with the Guide for the Care and Use of Laboratory Animals (National Research Council, National Academy Press, 2011). All experimental studies were performed on groups of clinically healthy rats, formed according to the method of analogs: same breed, same sex, same age, same weight. The animals were fed for 40 days, during the whole experiment the rats were kept in individual cells (10 animals each). Before introducing the studied fat into the diet, the animals were kept in quarantine for 21 days and transferred to an experimental diet in accordance with the experiment schedule.

The autopsy was performed with detailed logging and photographing of the material. Patho-morphological changes were studied on the material of 30 killed animals (Recommendations for euthanasia of experimental animals: Part 1, Part 2. [LaboratoryAnimals (1996) 30, 293-316; (1997) 31, 1-32]).

An autopsy was performed in the first 2 hours after killing. For histological examination, pieces of liver, spleen, and aorta were taken. To fix the pathological material, a 10% solution of neutral formalin was used.

Sections were obtained on a freezing microtome of model 2515 (Reichert Wien). The histological sections were stained with Ehrlich hematoxylin and eosin followed by microscopy. The morphological structure of the organs was studied in 30 fields of view of the microscope in different histological sections. Histological examination of the manufactured preparations was carried out under different magnifications, with detailed logging and photographing of the sites studied. Microphotography of histological specimens was performed using a CANON Power Shot A460 IS camera.

Clinical blood tests are conducted by apparatus method on haematologicalanalyser`s: PSE 90 Vet, Biochem SA (production of USA).

The characteristics of the diet and groups of animals are listed in table 1, the components of the diet and its nutritional value are in tables 15 and 16.

Table 1. Diet characteristics of the control and experimental groups of animals (Personal research data)

Reference number and the name of animal group	Diet	The energy value of the diet, kcal	The ratio of basic nutrients in the diet P:F:C *	The content of CIPE, %
Control group	Balanced diet	147	1:1:2	-
Experimental group N°1 (crackers)	Habitual balanced diet including crackers	140	1:1:2	1,32
Experimental group N°2 (*chak-chak*)	Habitual balanced diet including *chak-chak*	142	1:1:2	1,6
Experimentalgroup N° 3 (potato chips)	Habitual balanced diet including potato chips	140	1:1:2	2,55
Experimentalgroup N°4 (shortbread biscuits)	Habitual balanced diet including shortbread biscuits	160	1:1:2	3,65

*protein:fat:carbohydrates ratio

As can be seen from the table 1, the diet of all experimental groups of animals according to the ratio of basic nutrients corresponded to the norms of feeding experimental rats, the energy value fluctuated within ± 5%, which does not contradict the requirements for this indicator (Loskutova, Z.F., 1980; Skopichev, 2003)

While being daily monitored, the experimental groups of rats demonstrated worsening of the hairline, aggressiveness in the behavior by the middle of the experiment, which by the end of the experiment was replaced by apathy.

The most noticeable was experimental group No. 3, the animals of which were fed potato chips into the diet; in the middle of the experiment a surge of aggressiveness was observed, at the end the animals became passive and apathetic.

Pathological studies showed pallor and cyanosis of the mucous membranes, dullness of hair in rats of the experimental groups compared with the control group.
The results of the pathological studies are presented in table 2.

Table 2. The results of the pathological studies (Personal research data)

Indices	Control group	Experimental group N°1 (crackers)	Experimental group N°2 (*chak-chak*)	Experimentalgroup N° 3 (potato chips)	Experimentalgroup N°4 (shortbread biscuits)
State of the digestive organs	Nochangesdetected.	Granular focal liver dystrophy	Liverhyperemia	Granular diffuse and focal liver dystrophy	Focal granular dystrophy and liver hyperemia

The data in Table 2 indicates that in all experimental groups of rats, certain pathological processes at the tissue level were noted, with the greatest changes taking place in animals with a high content of CIPE in the diet - 3.65 and 2.55% at a normal rate of 1%.

Table 3. The results of histological studies (Personal research data)

Digestive organs	Control group	Experimental group N°1 (crackers)	Experimental group N°2 (*chak-chak*)	Experimentalgroup N° 3 (potato chips)	Experimentalgroup N°4 (shortbread biscuits)
1	2	5	6	3	4
Liver	The central veins of the lobules and sinusoidal hemocapillaries flowing into them are evenly expanded and filled with blood. The beam structure is preserved (Fig. 1). The boundaries of the liver cells are clear. The cores are well contoured.	Tissue elements unevenly perceived color. Perivascular lymphoid infiltration (Fig.3) Focal granular dystrophy.	Focal granular liver dystrophy. (Fig.5).	Tissue elements unevenly perceived color, the beam structure is poorly noticeable, hepatocytes in many fields of view of the microscope are arranged randomly. Diffuse granular and focal dystrophy (Fig.6). Diapedeshemorrhagesbetweenhepatocytes.	Perivascular edema (Fig. 8). Hyperemia, focal granular dystrophy.
Intestine	The structure of the intestinal wall is not disturbed, tinctorial properties are preserved, the villi of the mucous membrane are compact, not deformed, the vessels are moderately filled with blood (Fig. 2).	Catarrhal enteritis (Fig.4). Edema and lymphoid infiltration of the wall.	Desqumation of the glandular epithelium.	Edema of the mucous membrane (Fig. 7). Lymphoid infiltration and fragmentation of mucosal glands.	Edema of the mucous membrane. Perivascular edema, lymphoid infiltration of the glands, hyperplasia of the lymphoid follicle (Fig. 9). Lymphoid infiltration, mucous dystrophy of the glandular epithelium.

The results of histological studies confirm the presence of the changes. The histologic pattern is presented in table 3.

Figures 1 and 2 show micrographs of histological sections of the liver and intestines of the control group of animals.

Figure 1. Micrograph of a slice of the liver. Tinctorial properties saved. H&E staining x 150

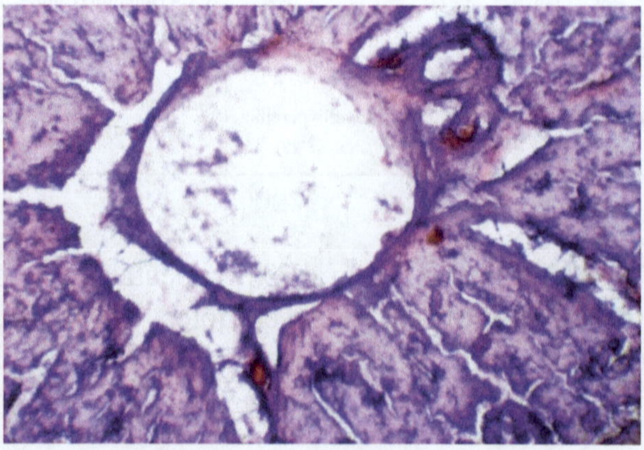

Figure 2. Micrograph of a slice of the intestine. Tissue structure unchanged. H&E staining x 50

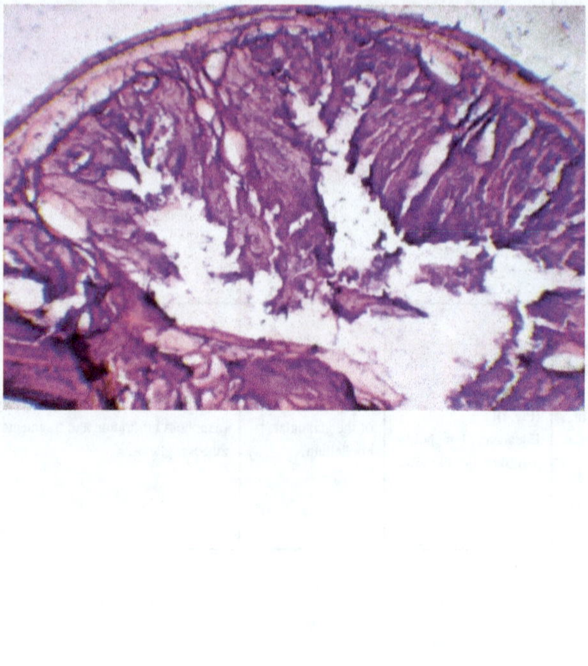

Figures 3-9 show micrographs of histological sections of experimental groups of animals. Experimental group N°1 (crackers).

Figure 3. Micrograph of a slice of the liver. Perivascular lymphoid infiltration. H & E staining x150

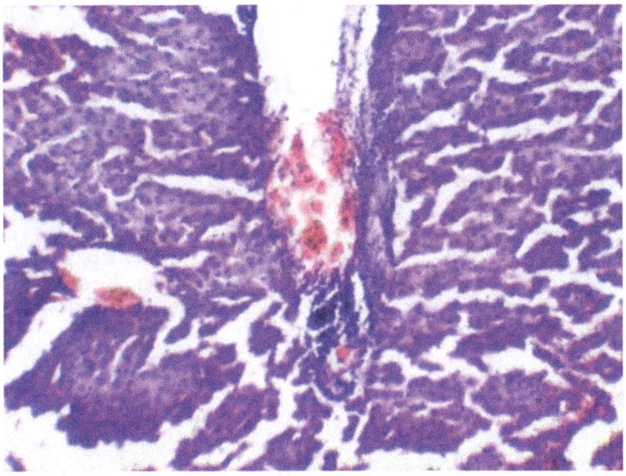

Figure 4. Micrograph of a slice of the intestine. Catarrhal enteritis. H & E staining x150 Experimental group N°2 (chak-chak)

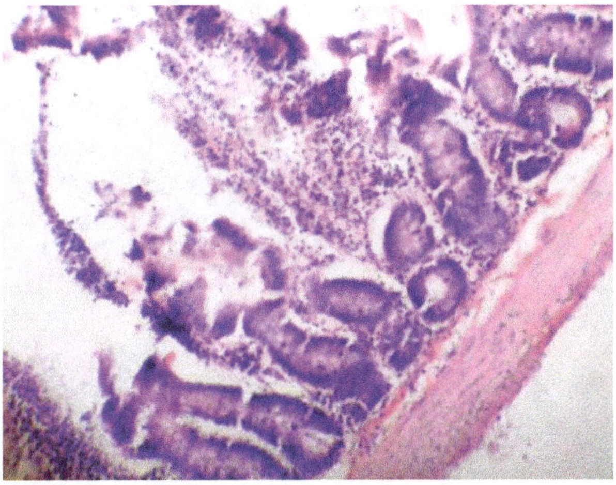

Figure 5. Micrograph of a slice of the liver. Granular dystrophy. H & E staining x 150Experimentalgroup Nº 3 (potato chips)

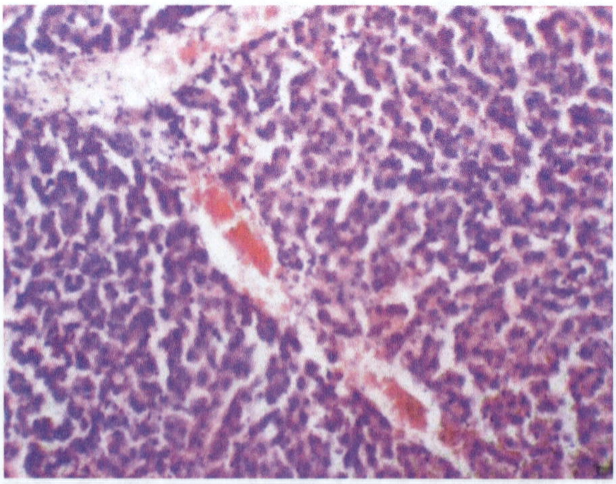

Figure 6. Micrograph of a slice of the liver. Diffuse granular dystrophy. H & E staining x150

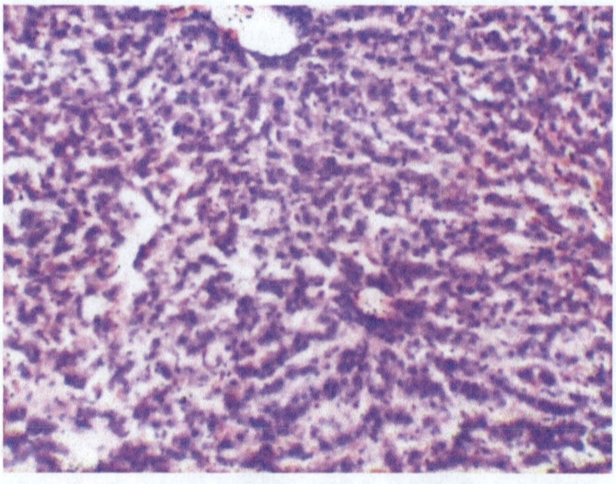

Figure 7. Micrograph of a slice of the intestine. Edema of the mucous membrane. H & E staining x150 Experimentalgroup N°4 (shortbread biscuits)

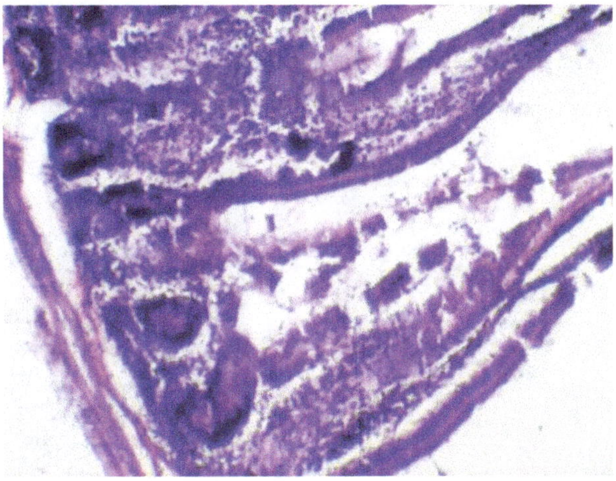

Figure 8. Micrograph of a slice of the liver. Perivascular edema. H & E staining x150

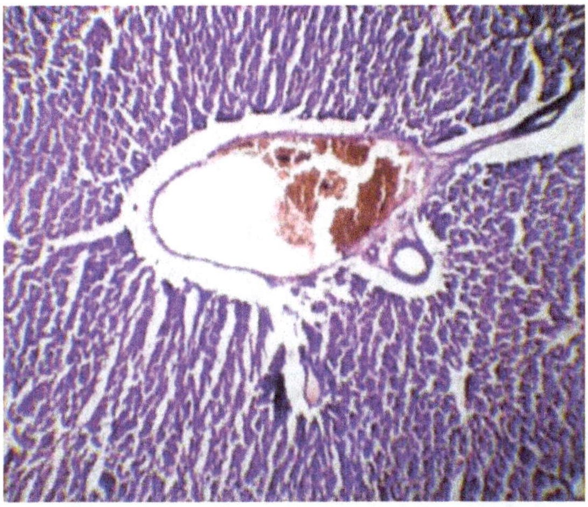

Figure 9. Micrograph of a slice of the intestine. Perivascular edema, lymphoid infiltration of glands, lymphoid follicle hyperplasia. H & E staining x50

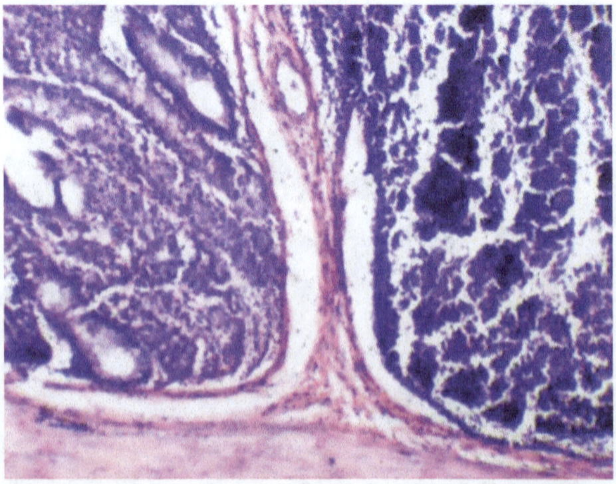

When comparing the results of histological studies, it is seen that in all animals from the experimental groups, in comparison with the control group, there are changes in the liver and small intestine with varying degrees of severity.

Analysis of the revealed changes has showed that with an increase in the concentration of compounds insoluble in petroleum ether (CIPE), the severity of pathological processes at the cellular level increases. In the control group of rats pathology has not been observed.

The most noticeable changes were found in rats of experimental groups No. 3 and 4, part of the daily ration of which was replaced by potato chips and shortbread biscuits. In these groups, there is diffuse granular and focal dystrophy of the liver, which causes the subsequent development of cirrhotic changes in the architecture of the liver. In the small intestine, edema of the mucous membrane was observed against the background of lymphoid infiltration and gland fragmentation.

In experimental groups of rats, in diets of which the content of CIPE was lower compared to the groups No. 3,4, but exceeded the permissible regulatory documents level of 1%, the changes were less noticeable, although they were found to be pathological processes in the liver and intestines. In group No. 1, which was receiving crackers, we found granular focal dystrophy of the liver, catarrhal enteritis, in addition, edema and lymphoid infiltration of the intestinal wall.

In group No. 2, where the part of the daily ration was replaced by a flour product chak-chak, diffuse granular dystrophy of the liver, changes in the intestine were found to be desquamation of glandular epithelium.

The results of pathological and histological studies indicate a negative effect on the body of products with an oxidized fatty component with prolonged consumption. The greater the concentration of CIPE is, the higher the intensity of pathological processes is.

In addition, we carried out hematological studies, since a blood test is an important component of the diagnosis of most diseases, and characterizes the general condition of the body (Tsynko, T.F., 2005). Blood is an internal medium that ensures the normal functioning of all organs and tissues of a living organism, reflecting to a greater or lesser extent the functional abnormality. The results of the general blood test are presented in table 4.

Table 4. Results of the general blood test of rats (Personal research data)

Indices	White rat (normogram, average)	Control group	Experimental group N°1 (crackers)	Experimental group N°2 (*chak-chak*)	Experimentalgroup N° 3 (potato chips)	Experimental group N°4 (short-bread biscuits)
1	2	3	4	5	6	7
Haemoglobin, GM/DL	150	145	82,0	71,0	83	49
Hematocrit,%	46	47,2	24,0	27,3	24,2	13,3
Erythrocytes, 10^{12}/l	7,5	7,6	4,4	3,98	4,61	2,8
Leukocytes, 10^{9}/l	10,5	14,6	3,4	6,6	3,8	3,3
Platelets, 10^{9}/l	400,0	610,5	142	135,0	192	177
ESR mm / hour	-	1	6	9	12	6
Basophils, %	0,5	0	1	1	0	1
Eosinophils, %	3,0	3	12	13	12	10
Juvenile,%	-	0	0	0	0	1
Bandnuclearcells, %	2,0	4	4	6	5	4
Segmental, %	25,5	32	12	4	17	38
Lymphocytes, %	65,0	56	64	73	62	41
Monocytes, %	3,0	3	6	3	4	5
Anisocytosis,%	-	10,4	14,2	12,5	14,0	15,7
Colorindicator	0,8	0,9	0,6	0,6	0,7	0,6

The data in Table 4 indicate a change in the morphological composition of the blood, expressed in the phenomena of severe leukopenia, erythropenia, which, in combination with a decrease in hemoglobin, resulted in anemia, thrombocytopenia, a significant increase in ESR. In the leukogram of animals of group No. 4., the changes were represented by monocytosis, eosinophilia and lymphocytopenia.

A sharp decrease in the level of leukocytes and erythrocytes in all experimental groups of rats indicates that the oxidation products of fats demonstrate the properties of leukotoxins and have a negative effect on the concentration of erythrocytes. These results have been obtained for the first time and partially correspond to some research results by foreign scientists, in particular, Markaverich, B.M. and Ma´rquez-Ruiz G. (Velasco, J. et al., 2004; Markaverich B.M. et al., 2005; Marquez-Ruiz, G. et al., 2006).

The results of biochemical analysis (Table 5) also indicate negative changes in the body during the consumption of products containing a significant amount of toxic products of oxidation of fats.

Table 5. Biochemical blood parameters of rats (Personal research data)

Indices	White rat (normogram, average)	Control group	Experimental group N°1 (crackers)	Experimental group N°2 (*chak-chak*)	Experimentalgroup N° 3 (potato chips)	Experimentalgroup N°4 (shortbread biscuits)
Totalbilirubin, mcmole/l	2,9	5,1	14,5	7,8	17,5	16,5
Cholesterol, mol / l	2,5	2,5	8,6	7,0	9,3	8,6
Totalprotein GM/DL	67,0	66,0	45,0	52,5	36,2	41,7
Creatinine, mol / l	95,0	95,0	83,7	68,3	63,4	73,0
Amylase, units / l	1150,0	1147,0	1790,0	1100,5	1750,0	1627,5

Biochemical blood tests (table 5) showed significant changes in the level of bilirubin, i.e. an increase by 2 or more times, cholesterol, i.e. an increase by more than 70-80% and a significant decrease in the level of protein, i.e. by 30-40%.

In this case, the most serious deviations from the norm are observed in the experimental group of rats No. 3, which were fed with potato chips of industrial production.

Thus, all the changes in the groups of rats that consumed fast food products of industrial production with the oxidized fat phase demonstrated inhibition of hematopoietic activity, immunodeficiency and inflammatory process in the acute form. The greater the concentration of compounds insoluble in petroleum ether (CIPE) in the diet is, the more intense the changes are.

As a result of the experiments, it has been first discovered that the modified fatty component of fast food products, containing more than 1% of CIPE, regardless of its production technology (deep-frying, baking, drying), equally affects the body. This makes it possible to correlate the degree of negative influence and the level of concentration of CIPE and recommend rationing this indicator in fast food products with a large proportion of the fatty component intended for long-term storage.

It should be noted that the products used in the experiment had different fatty acid composition, which could affect the objectivity of the results of experimental studies. In order to reliably indicate the dependence of pathological changes in the body on the concentration of CIPE, it is advisable to conduct research on one type of fatty component with different content of CIPE in the diet.

7.2 Investigation of The Intensity of Pathogenesis Dependingon The Concentration of Cipe In Palm Oil

Studies were conducted using palm oil of the same batch, heat treated in the range from one to 5 days and containing various levels of CIPE.

Characteristics of diets and groups of experimental animals are presented in Table 6.

Table 6. Characteristics of the diet of experimental groups of animals (Personal research data)

Reference number and the name of animal group	Diet	The energy value of the diet, kcal	The ratio of basic nutrients in the diet P:F:C [173]	The content of CIPE, %
Control group	Habitual full diet with the inclusion of the original palm oil	132	1:1:2	0,17
Experimental groupN°1	Habitual full diet with the inclusion of thermally oxidized palm oil	132	1:1:2	0,88
Experimental group N°2		132	1:1:2	1,8
Experimentalgroup N° 3		132	1:1:2	2,89

For the accuracy of the experiment, animals that received palm oil containing up to 0.2% of CIPE were used as a control group in these studies.

During external examination of involuntarily killed rats, no changes were found.

When comparing the results of histological studies (table 7) in the liver and intestines in rats in all groups, pathological changes of different severity were found. In the control group that received the original palm oil, there were no changes in the liver, stomach, small intestine, and only in the large intestine an increase in the number of goblet cells and edema in the submucosal layer was detected.

The histological picture in group 1 in comparison with the control group indicates the presence of small changes in all the digestive organs (table 7).

Table 7. The results of histological examination (Personal research data)

Digestive organs	Control group	Experimental group N°1	Experimental group N°2	Experimentalgroup N° 3
	CIPE, %			
	0,17%	0,88%	1,8%	2,98%
1	2	3	4	5
Liver	The structure of the organ is within the physiological norm.	Hyperemia, granular dystrophy of hepatocytes.	Congestive hyperemia, granular dystrophy, karyolysis, desquamation of the epithelium of the bile ducts.	Congestive hyperemia, granular dystrophy, karyolysis.
Gaster	The structure of the organ is within the physiological norm.	Edema of the submucous and muscular layer, edema of the integumentary epithelium.	Desquamation of the integumentary epithelium, hyperemia, edema of the submucosal layer, dystrophy of muscle fibers.	Mucous dystrophy of the integumentary epithelium, desquamation of the epithelium, lymphoid infiltration of the submucosal layer.
Small intestine	The structure of the organ is within the physiological norm.	Hyperemia and hemorrhage, infiltration of the muscular layer by lymphoid cells, desquamation of the integumentary epithelium, an increase in goblet cells.	An increase in goblet cells, hyperemia, lymphoid infiltration of the muscular layer.	Marked mucous dystrophy, desquamation of the epithelium, hyperemia, lymphoid infiltration.
Large intestine	Slight edema of the submucous layer, an increase in goblet cells.	Hyperemia, edema, increase of goblet cells.	Desquamation of the integumentary epithelium, an increase in lymphatic follicles, edema and hyperemia.	Hemorrhages, mucous dystrophy, desquamation of the integumentary epithelium, lymphoid infiltration of the submucosal layer.

More intense changes in the organs of the digestive system were observed in group No. 2, where rats received palm oil, thermally oxidized for 3 days. Changes in the liver are characterized by an increase in the size of hepatocytes with the appearance of protein-like granules in their cytoplasm. In the stomach, there is edema of the submucosal layer, the appearance of areas of muscle dissociation. Changes in the duodenum and large intestine were also detected.

The most significant changes can be seen in group No. 3 which received thermo-oxidized palm oil in the diet, which was heated for 5 days. In rats of this group, in separate areas, the processes of beam structure uncomplexing and an increase in the volume of hepatocytes, were detected. In the stomach, the epithelium is in a state of mucous dystrophy and desquamation, and in the muscle layer there is edema and dissociation.

In the duodenum, noticeable mucosal dystrophy with hypersecretion of mucus, necrosis and desquamation of epithelial cells were detected. In the submucosal layer of the large intestine, there is dissociation of the connective tissue elements, the epithelium of the mucosa is in a state of mucosal dystrophy and desquamation.

Studies have shown that thermally oxidized palm oil, which was heat treated for 1-5 days, as the oxidative destruction products accumulate in it, adversely affects the body, causing pathological changes in the digestive system. The analysis of the results of our experimental study allows us to establish a close causal relationship between the content of compounds insoluble in petroleum ether (CIPE) and the effect of thermally oxidized palm oil on the body. The pathological changes in the internal organs begin as early as the content of secondary oxidation products in it reaches the level of 0, 88%, which is very close to the current rate of 1%.

These experiments lead to a scientifically based conclusion that the current standard for the content of secondary oxidation products is quite adequate, but in the future it is necessary to tighten the requirements for the level of oxidized compounds in palm oil during the production of deep-frying fat products.

7.3 Comparative Analysis of The Level of Toxic Effects of Deep-Frying Fats and Fast Food Products Fried in Them

At the next stage of research, we carried out a comparative analysis of the level of toxic effects which the consumption of deep-frying fats and fast food products, fried in them, and containing different concentrations of CIPE, has on the body. The studies are conducted due to the lack of data on the degree and comparability of the effect of deep-frying fats and products, fried in them, on the body.

A number of research papers have shown that the products being fried adsorb oxidation products from deep-frying fat, thereby reducing their content in fat (Shilman, L.Z., 1966; Terzieva, V., 1990; Velasco, J. et al., 2004). Greater attention is given to the data that the fat extracted from the product and deep-frying fat are oxidized to the same degree (Kravchuk, N.N., 1981; Nosova, AS, 2013).

Our research has shown that the fat contained in the product is obtained mainly as a result of the absorption of deep-frying fat by finished fast food products, so that the deep-frying fat and the fat extracted from the product are approximately similar in terms of the content of oxidation products.

When conducting experiments, the same fat concentration in the diet was maintained when deep-frying fat and finished product were consumed (table 76). Experimental groups of animals had a balanced experimental diet, in which a part of food items was replaced with the samples of fats and fast food products. Characteristics of diets and groups of experimental animals are presented in table 8.

Table 8. Characteristics of the diet of experimental groups of animals (Personal research data)

<table>
<tr><th>Reference number and the name of animal group</th><th>Diet</th><th>The energy value of the diet, kcal</th><th>The ratio of basic nutrients in the diet P:F:C [14]</th><th>The content of CIPE, %</th></tr>
<tr><td>Control group</td><td>Habitual full diet [14]</td><td>147</td><td>1:1:2</td><td>-</td></tr>
<tr><td>Experimental group N°1</td><td>Habitual full diet with the inclusion of chak-chak</td><td>142</td><td>1:1:2</td><td rowspan="2">1,67</td></tr>
<tr><td>Experimental group N°2</td><td>Habitual full diet with the inclusion of fat after frying chak-chak</td><td>132</td><td>1:1:2</td></tr>
<tr><td>Experimental group N°3</td><td>Habitual full diet with the inclusion of french fries</td><td>143</td><td>1:1:2</td><td rowspan="2">1,92</td></tr>
<tr><td>Experimental group N°4</td><td>Habitual full diet with the inclusion of fat after frying french fries</td><td>132</td><td>1:1:2</td></tr>
</table>

Table 9 presents analytical data showing the behavior of animals during the experiment.

Table 9. Behavioral characteristics of experimental groups of rats during the experiment (Personal research data)

Experimental group of rats	The behavior of experimental groups of rats			
	Week 1	Week 2	Week 3	Week 4
Control	Active	Active	Active	Active
Group 1	Active	Active	Frequent mood swings	Passive
Group 2	Active	Frequent mood swings	Passive	Refusal from food
Group 3	Active	Frequent mood swings	Passive	Refusal from food
Group 4	Active	Passive	Passive	Refusal from food

Analyzing the data of table 9, we can conclude that the products included in the diet of animals, have an impact on their behavioral aspects.

In particular, the behavior of the animals of the experimental groups at the beginning of the experiment did not differ from animals of the control group. From the second week of the experiment, frequent mood swings were observed; by the end of the experiment, the animals were passive, and the food was poorly eaten.

The results of the pathological study are presented in table 10.

Table 10. The results of the pathological study of rats after feeding them deep-frying fats and foods fried in them (Personal research data)

White rat	State of the digestive organs
Control group	Norm
Group 1 (chak-chak)	Liver hyperemia
Group 2 (fat after frying chak-chak)	Liverhyperemia
Group 3 (french fries)	Focalliverdegeneration
Group 4 (fat after frying french fries)	Focalliverdegeneration

The analysis of data from a pathological study shows that changes in the liver occurred in all experimental groups of rats. They were of a similar nature in groups where in diets deep-frying fats and foods fried in them were introduced.

No changes at the tissue and cellular levels were detected in the control group of rats.

Histological studies have shown that the products (chak-chak, french fries) that were used in the experiment, have a toxic effect even against the background of a balanced diet, apparently as a result of the adsorption of deep-frying fat containing toxic oxidation products. This effect appears in the development of such changes as granular dystrophy of the liver, hyperemia, as well as desquamative and edematous processes in the intestinal mucosa (Figures 10-14).

Figure 10. Micrograph of a section of the liver, hyperemia. A group of rats was fed with deep-frying fat, after frying Chak-chak. H & E stainingx 50

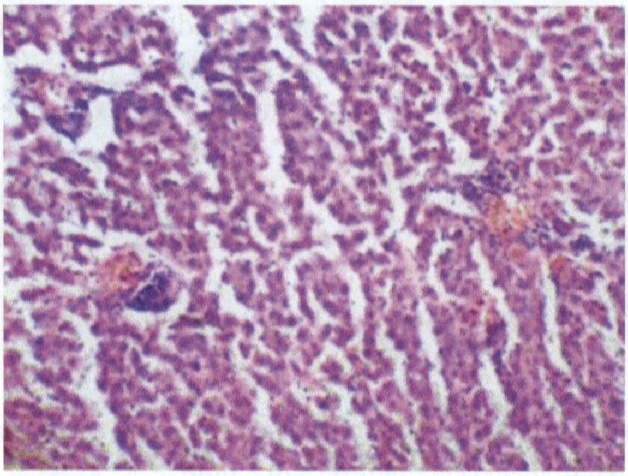

Figure 11. Micrograph of intestinal section, desquamation of glandular epithelium. A group of rats was fed with Chak-chak. H & E stainingx 50

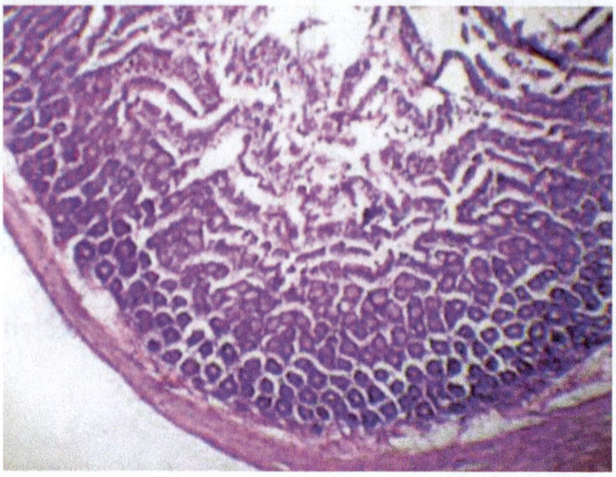

Figure 12. Micrograph of a section of the liver, perivascular edema. A group of rats was fed with potato cut into cubes and deep-fried. H & E stainingx 150

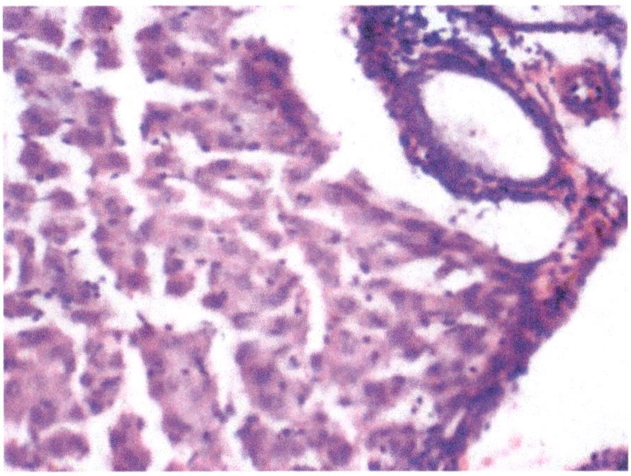

Figure 13. Micrograph of the intestinal section - desquamation of the glandular epithelium. A group of rats that was fed with deep-frying fat after frying potatoes. H & E stainingx 50

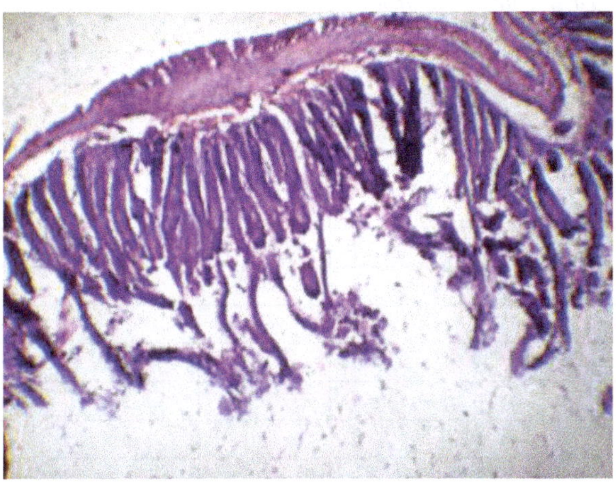

Figure 14. Micrograph of a slice of the intestine - wall edema. A group of rats was fed potatoes cut into cubes and deep-fried. H & E stainingx 50

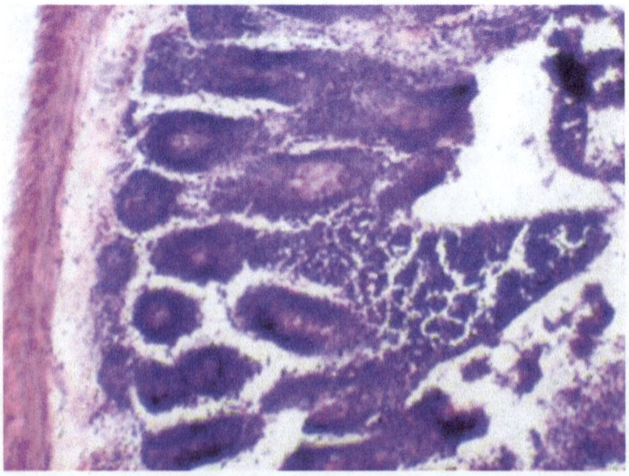

A similar histological pattern was obtained in the experimental groups of rats, which were fed with deep-frying fats used to fry chak-chak flour product and french fries (Figures 10, 13).

Table 11. Results of the general blood test (Personal research data)

Indices	White rat (normogram, average)	Control group	Chak-chak (Experimental group 1)	Fat after frying chak-chak (Experimental group 2)	French fries (Experimental group 3)	Fat after frying french fries (Experimental group 4)
Haemoglobin, GM/DL	150	145,0	71,0	71	67,0	56
Hematocrit, %	46	31,4	27,3	20,6	47,2	16,5
Erythrocytes, 10^{12}/l	7,5	7,6	3,98	3,98	4,87	2,82
Leukocytes, 10^9/l	10,5	14,6	6,6	6,6	3,6	3,7
Platelets, 10^9/l	400,0	132,0	135,0	180	610,5	90
ESR mm / hour	-	12,0	9	12	1	12
Basophils, %	0,5	0	1	0	0	3
Eosinophils, %	3,0	3,0	13	13	13	10
Myelocytes,%	-	0	0	0	0	0
Juvenile,%	-	0	0	0	0	0
Bandnuclearcells, %	2,0	3	6	6	6	5
Segmental, %	26,5	17	4	5	32	3
Lymphocytes, %	65,0	69	73	72	56	76
Monocytes, %	3,0	4	3	4	3	3
Anisocytosis, %	-	13,9	12,5	13,0	10,4	15,6
Colorindicator	0,8	1,1	0,6	0,6	0,6	0,5

At the next stage, hematological studies were performed. Table 11 presents the results of the general blood test.

The changes presented in table 11 were of diverse nature. Thus, in animals of the second and fourth experimental groups, we observed a decrease in the level of hemoglobin, erythrocytes, and platelets with a strong increase in ESR and the phenomenon of eosinophilia

Changes in the experimental groups of rats that consumed chak-chak and french fries, were characterized by a deviation from the physiological norm of hemoglobin, the decrease of erythrocyte and leukocyte levels and the presence of eosinophilia, which was common to all experimental groups. The number of platelets in group 1 was significantly reduced compared with the control group together with an increase in ESR. In group 3, the number of platelets increased while reducing ESR.

Table 12 illustrates the biochemical blood parameters of animals.

Table 12. Biochemical blood parameters of rats (Personal research data)

Indices	White rat (normogram)	Control group	Experimental group 1 Chak-chak	Experimental group 2 Fat after frying chak-chak	Experimental group 3 french fries	Experimental group 4 Fat after frying french fries
	average					
Totalbilirubin, mcmole/l	7,2	7,1	7,8	19,5	13,4	13,2
Cholesterol, mol / l	4,4	4,5	7,0	9,6	7,9	7,5
Totalprotein GM/DL	67	66	52,5	26,5	34,5	54,0
Creatinine, mol / l	95	95	68,3	93,0	65,0	67,3
Amylase, units / l	1150	1147	1100,5	2105,4	1640,5	1410,5

Changes in the biochemical composition appeared with hyperbilirubinemia and hypercholesterolemia against in conjunction with general hypoproteinemia. A sharp increase in the amount of amylase by 2 or more times was also detected.

Thus, in all experimental groups of rats, leukopenia, inhibition of hematopoietic activity, immunodeficiency states and inflammatory processes in different degrees of exacerbation were observed. The greatest changes occurred in the body of the experimental groups of rats fed with fat for frying chak-chak and deep-fried potatoes (french fries).

We assume, the most significant is a sharp decrease in the content of leukocytes and erythrocytes, which indicates the presence of compounds that are leukotoxins in deep-frying fat and fast food products (Velasco, J. et al., 2004; Markaverich B.M. et al., 2005; Marquez-Ruiz, G. et al., 2006)., and some factors affecting the content of red blood cells. The mechanisms of influence of toxic oxidation products on the body are not fully understood.

The conducted studies revealed for the first time an identical toxic effect that the fats and products fried in them have on the body.

In view of these studies and the severity of the physiological consequences, since deep-fried products and the fats used to produce them can be categorized as chemical hazards, the requirements for the safety of deep-fried products and the fats used to produce them are of great importance, but so far insufficiently regulated by any regulatory document.

Malignant tumors are the second most common cause (after cardiovascular disease) of human morbidity and mortality.

Every year in industrialized countries, the number of deaths from malignant neoplasms reaches 30% of all deaths caused by diseases, and these indicators continue to grow steadily, as fast food consumption continues to grow. As a result, it is necessary to estimate the importance of controlling the technological process of producing fast food products, which manufacturers often neglect, not complying with established hygienic standards for the content of oxidation products insoluble in petroleum ether (CIPE), focusing on other indicators that do not correlate with the safety of deep-frying fat, for example, such as the content of free fatty acids or organoleptic indicators of deep-frying fat. It should be noted that the test samples of deep-frying fat and fast food products were distinguished by positive organoleptic characteristics. The use of deep-frying fat in the production process was discontinued according to the results of instrumental control of the products of fat oxidation even before the onset of unsatisfactory organoleptic evaluation.

7.4 Safety Assessment of Special Deep-Frying Fats in Animal Experiments

In subsequent clinical experiments, the intensity of pathological processes in the cardiovascular and excretory systems was investigated, depending on the concentration of oxidation products in the deep-frying fat of the new generation Vegafray 05, which does not contain trans-isomers (Cargill Russia LLC), after high-temperature heating. The basis for the study of this fat was the fact that this type of fat is widely used in fast food chains, such as McDonalds, KFC, Burger King, and Irish Pub, and, in addition, the fat-acid composition of this fat is close to optimum.

Vegafray 05 fat contains an antioxidant complex (E319 tert-butylhydroquinone, complexing agent E330 citric acid, E900 polydimethylsiloxane defoaming agent), which is added to fat as a solution in pure propylene glycol (E 1520). Food supplements used in amounts permitted provide fairly good technological characteristics of the produced deep-frying fats.

The diet description of experimental groups of animals is presented in Table 13.

Daily observation of the animals showed that all the experimental groups of animals showed deterioration in appetite and appearance, condition of the hair coat. The experimental group No. 3 stands out the most, as the animals were given thermally oxidized deep-frying fat with a content of secondary oxidation products (CIPE) of 1%. In this group, by the end of the experiment, a surge of aggressiveness was observed.

Table 13. Diet of experimental animals (Personal research data)

Reference number and the name of animal group	Diet	The energy value of the diet, kcal	The ratio of basic nutrients in the diet P:F:C [14]	The content of CIPE, %
Control group	Habitual full diet	147	1:1:2	
Experimental group N°1	Habitual full diet with the inclusion of thermally oxidized fat "Vegafray 05"	132	1:1:2	0,42
Experimental group N°2		132	1:1:2	1,04

Pathological studies in the control group of rats showed no abnormalities; in the experimental groups of rats, changes in the liver, kidneys, and cardiovascular system were observed, represented by hyperemia of the liver, diffuse dystrophy of the kidneys, and vascular hyperemia.

The results of histological studies are presented in table 14 and figures 15-22.

Table 14. The results of histological studies (Personal research data)

Organs	Control group	Experimental group N°1	Experimental group N°2
Myocardium	The structure of the organ at the cellular level has not been changed.	Marked perivascular edema and adventitia dissection	**Pathology:** marked edema of the aortic wall, edema of the vascular endothelium
Liver	The structure of the organ at the cellular level has not been changed.	**Pathology:** severe diffuse granular dystrophy, diapedemic hemorrhages	**Pathology:** granular dystrophy of hepatocytes, organ tissue edema. Multiplediffusediapedemichemorrhagesandvascularhyperemia
Kidneys	The structure of the organ at the cellular level has not been changed.	**Pathology:** hyperemia of blood vessels, diffuse granular dystrophy of the epithelium of the tubules, serous glomerulitis, marked hyperemia of the glomerular vessels, focal diapezous hemorrhages	**Pathology:** vascular hyperemia, diffuse granular dystrophy, perivascular edema, serous glomerulitis
Smallintestine	The structure of the organ at the cellular level has not been changed.	**Pathology:** edema, deformity and rejection of the glands	**Pathology:** marked edema of the mucous membrane. Deformation and destruction of the glands. Focalparabioticprocessesinthemucousmembrane
Large intestine	The structure of the organ at the cellular level has not been changed.	**Pathology:** edema of the mucous membrane, destruction of glands	**Pathology:** marked edema of all layers of the mucous membrane, hemorrhage

In the control group, the histological pattern of these organs is normal, and there are no changes.

Experimental groups that consumed oxidized deep-frying fat in the diet were characterized by perivascular myocardial edema, diffuse degeneration of hepatocytes, impaired tinctorial properties of the kidney tissue, marked edema of the mucous membrane of the small and large intestine.
Experimentalgroup N°1

Figure 15. Micrograph of myocardial slice, aorta 2. Marked perivascular edema and stratification of adventitia. H & E stainingx 50

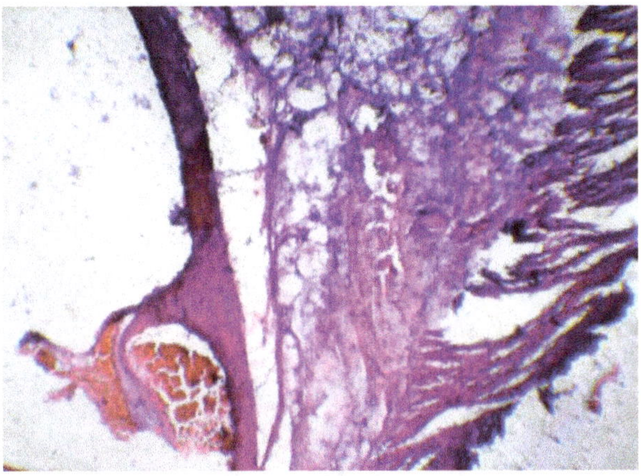

Figure 16. Micrograph of a section of the liver 2. Marked diffuse granular dystrophy. Perivascular edema. H & E stainingx 150

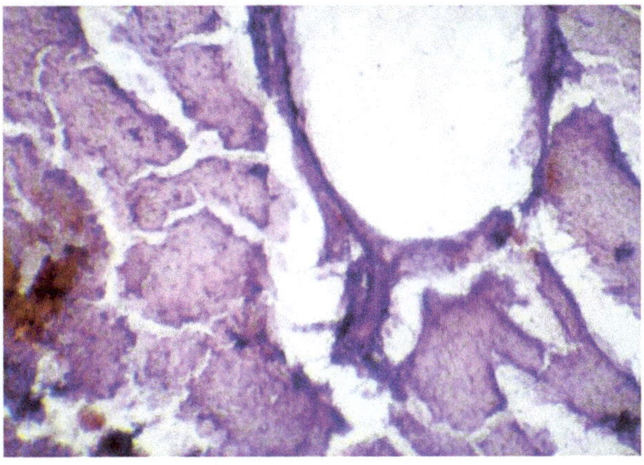

Figure 17. Micrograph of a section of the kidney 2. Serous glomerulitis, severe glomerular vascular hyperemia, focal diapedemic hemorrhages. H & E stainingx 300

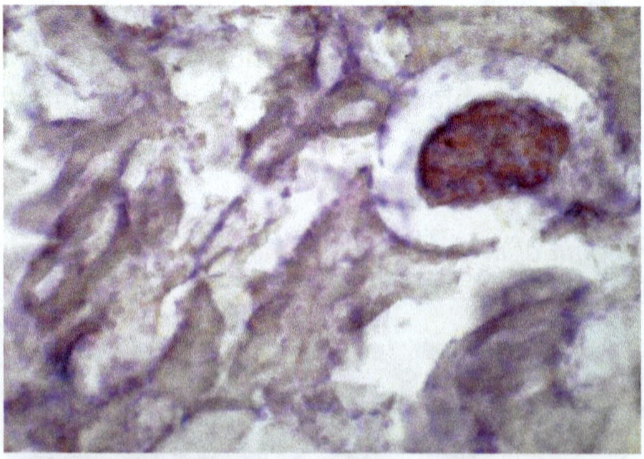

Figure 18. Micrograph of a slice of the large intestine 2. Damage of the mucosal glandular structures. H & E stainingx 150

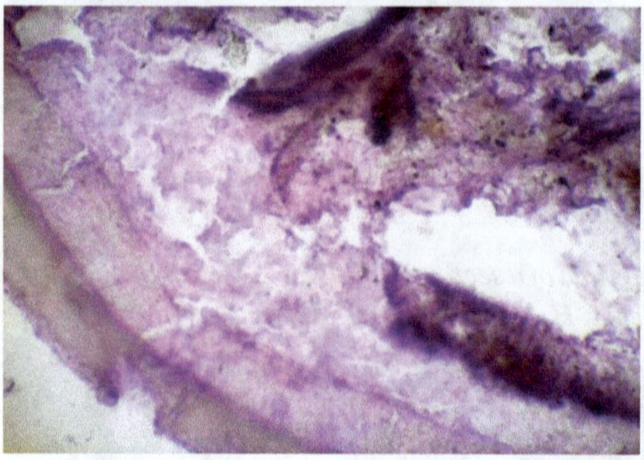

Experimental group N°2

Figure 19. Micrograph of a section of the liver 3. Multiple diffuse diapedemic hemorrhages and hyperemia of microvasculature vessels. H & E stainingx 300

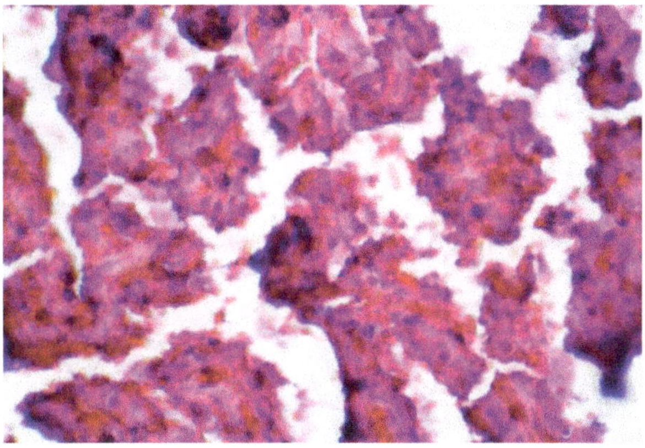

Figure 20. Micrograph of myocardial slice, aorta 3. marked edema in the muscle tissue and at the wall of the aorta. H & E stainingx 50

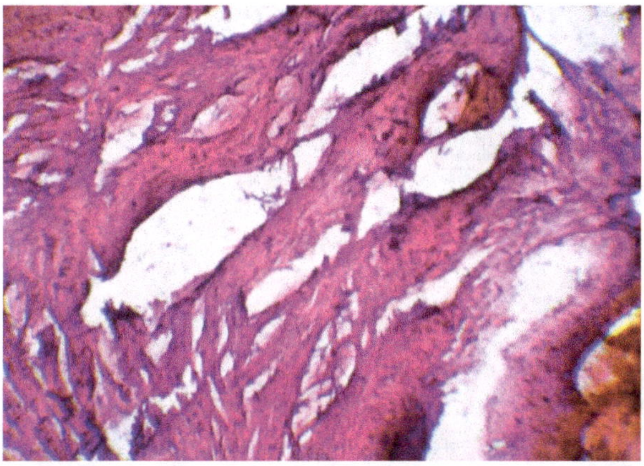

Figure 21. Micrograph of a slice of the small intestine 3. Marked edema of the mucous membrane. H & E stainingx 50

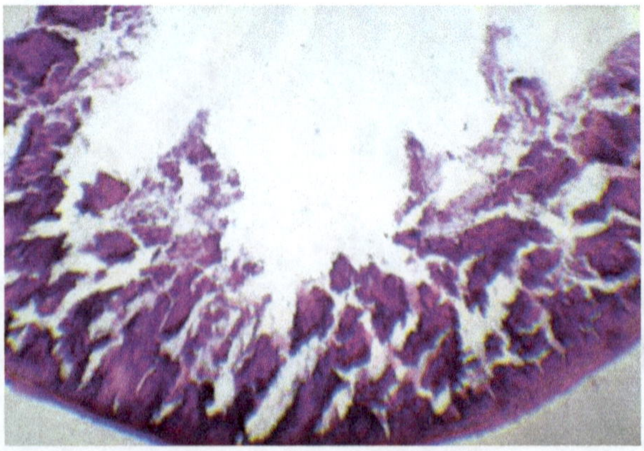

Figure 22. Micrograph of a section of a kidney 3. Serous glomerulitis, vascular hyperemia glomerulus and capillaries. H & E stainingx 300

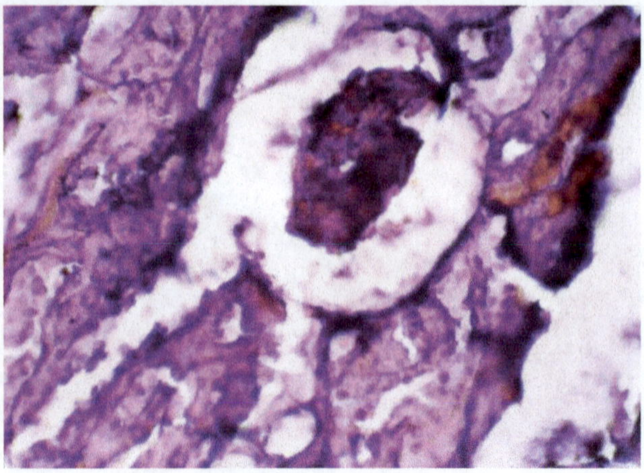

Studies have shown that changes in the organs of rats of the experimental groups were typical of a little intense, but long-lasting intoxication. It resulted in a gradual deterioration of the liver, an increase in the enzymatic processes in the gastrointestinal tract and, as a result, an increase in the load on the urinary and cardiovascular systems.

The difference in the groups was characterized only by the intensity of these pathological processes, depending on the content of CIPE in the diet.

Table 15. Content of basic elements in the diet of laboratory animals (Personal research data)

Food items1	Net weight, g2	Proteins, g3	Fats, g4	Carbohydrates, g5	Energy value, kcal6
Control group					
Mixture of cereals	15	2,1345	3,51	3,0525	147
Bread	4	0,976	0,488	3,332	
Oatmeal	3	0,729	0,147	0,4512	
Milk	8	0,28	0,28	1	
Meat	5	1,25	0,75	0	
Juicy foods	20	2,41	0,5	7,5	
Fish oil	0,1	0	0,0899	0	
Beefish flour	0,7	0,0039	0,00021	0	
Total		8,68	5,76	15,33	
Experimental groups					
French fries consuming group					
Mixture of cereals	15	2,1345	3,51	3,0525	143
Oatmeal	3	0,729	0,147	0,4512	
French fries	4	0,15	2,1	0,6	
Milk	8	0,28	0,28	1	
Meat	5	1,25	0,75	0	
Juicy foods	10	1,2	0,25	7,5	
Fish oil	0,1	0	0,0899	0	
Beefish flour	0,7	0,0039	0,00021	0	
Total		6,9	7,3	12,6	
1	2	3	4	5	6
Chak-chak consuming group					
Mixture of cereals	15	2,1345	3,51	3,0525	142
Outmeal	3	0,729	0,147	0,4512	
Chak-chak	4	0,47	0,8	2,6	
Milk	8	0,28	0,28	1	
Meat	5	1,25	0,75	0	
Juicy foods	20	2,41	0,5	7,5	
Fish oil	0,1	0	0,0899	0	
Beefish flour	0,7	0,0039	0,00021	0	
Total		7,2	6,07	14,6	
Used deep-frying fat consuming group					

Continued on following page

Table 15. Continued

Food items1	Net weight, g2	Proteins, g3	Fats, g4	Carbohydrates, g5	Energy value, kcal6
Mixture of cereals	15	2,1345	3,51	3,0525	132
Oatmeal	3	0,729	0,147	0,4512	
Used deep-frying fat	2	0	1,95	0	
Milk	8	0,28	0,28	1	
Meat	5	1,25	0,75	0	
Juicy foods	10	1,2	0,25	7,5	
Beefish flour	0,7	0,0039	0,00021	0	
Total		5,59	6,8	12,03	
Industrially produced crackers consuming group					
Mixture of cereals	15	2,1345	3,51	3,0525	140
Outmeal	3	0,729	0,147	0,4512	
Crackers	4	0,48	0,25	3,4	
Milk	8	0,28	0,28	1	
Meat	5	1,25	0,75	0	
Juicy foods	20	2,41	0,5	7,5	
Fish oil	0,1	0	0,0899	0	
Beefish flour	0,7	0,0039	0,00021	0	
Total		7,2	5,5	15,4	
Potato chips consuming group					
Mixture of cereals	15	2,1345	3,51	3,0525	140
Oatmeal	3	0,729	0,147	0,4512	
Potato chips	4	0,15	1,4	1,2	
Milk	8	0,28	0,28	1	
Meat	5	1,25	0,75	0	
Juicy foods	10	1,2	0,25	7,5	
Fish oil	0,1	0	0,0899	0	
Beefish flour	0,7	0,0039	0,00021	0	
Total		6,95	6,6	13,0	
Industrially produced shortbread biscuits consuming group					
Mixture of cereals	15	2,1345	3,51	3,0525	161
Outmeal	3	0,729	0,147	0,4512	
Shortbread biscuits	4	0,49	1,93	5,16	
Milk	8	0,28	0,28	1	
Meat	5	1,25	0,75	0	

Continued on following page

Table 15. Continued

Food items1	Net weight, g2	Proteins, g3	Fats, g4	Carbohydrates, g5	Energy value, kcal6
Juicy foods	20	2,41	0,5	7,5	
Fish oil	0,1	0	0,0899	0	
Beefish flour	0,7	0,0039	0,00021	0	
Total		7,29	7,2	17,1	
Crackers with antioxidants consuming group					
Mixture of cereals	15	2,1345	3,51	3,0525	
Oatmeal	3	0,729	0,147	0,4512	
Crackers	4	0,48	0,25	3,4	
Milk	8	0,28	0,28	1	
Meat	5	1,25	0,75	0	139,9
Juicy foods	10	1,2	0,25	7,5	
Fish oil	0,1	0	0,0899	0	
Beefish flour	0,7	0,0039	0,00021	0	
Total		7,2	5,5	15,4	
Potato chips with antioxidants consuming group					
Mixture of cereals	15	2,1345	3,51	3,0525	
Outmeal	3	0,729	0,147	0,4512	140
Potato chips	4	0,15	1,4	1,2	
Milk	8	0,28	0,28	1	
Meat	5	1,25	0,75	0	
Juicy foods	20	2,41	0,5	7,5	
Fish oil	0,1	0	0,0899	0	
Beefish flour	0,7	0,0039	0,00021	0	
Total		6,95	6,6	13,0	
Shortbread biscuits with antioxidants consuming group					
Mixture of cereals	15	2,1345	3,51	3,0525	
Oatmeal	3	0,729	0,147	0,4512	
Shortbread biscuits	4	0,36	0,75	2,15	
Milk	8	0,28	0,28	1	
Meat	5	1,25	0,75	0	139
Juicy foods	10	1,2	0,25	7,5	
Fish oil	0,1	0	0,0899	0	
Beefish flour	0,7	0,0039	0,00021	0	
Total		7,16	6,0	14,1	
Used and purified deep-frying fat consuming group					

Continued on following page

Table 15. Continued

Food items1	Net weight, g2	Proteins, g3	Fats, g4	Carbohydrates, g5	Energy value, kcal6
Mixture of cereals	15	2,1345	3,51	3,0525	132
Outmeal	3	0,729	0,147	0,4512	
Deep-frying fat after purification	2	0	1,95	0	
Milk	8	0,28	0,28	1	
Meat	5	1,25	0,75	0	
Juicy foods	20	2,41	0,5	7,5	
Beefish flour	0,7	0,0039	0,00021	0	
Total		5,59	6,8	12,03	

The analysis of the results of our studies showed that the consumption of thermally oxidized Vegafray 05 deep-frying fat, despite the optimum fatty acid composition, leads to impaired physiological processes in the body, even at relatively low concentrations of CIPE. It should be noted that the antioxidants, that are present in it, may also affect the toxicity of the fat used.

The components of the diet and its nutritional value are listed in the final tables of this chapter 15 and 16.

The content of vitamins and minerals is shown in Table 16. Vitamins and minerals were introduced into the food in the form of a premix.

Table 17. The content of vitamins and minerals in the dietof laboratory animals (Personal research data)

Vitamins mg				Minerals mg			
A	D	E	K	Fe	Mg	K	Ca
0,9	3,0	0,6	2,0	0,25	1,0	12,0	40,0

CONCLUSION OF THE CHAPTER

The analysis of the results of a series of experiments conducted on more than 200 experimental animals (white rats) with a study of various internal organs: liver, large and small intestine, stomach, kidney and myocardium, with a high degree of accuracy showed the negative physiological effects of products of fat oxidation on the body. All changes in groups of rats that consumed industrially produced fast-food products with oxidized fat phase and fats used for frying were a manifestation of inhibition of hematopoietic activity, hemodynamic and dystrophic disorders, immunodeficiency and inflammatory processes of varying intensity.

Such diverse experiments on animals were carried out for the first time, including not only histological studies that are considered classical, but also hematological studies, which revealed that, when ingested, the products of fat oxidation cause serious changes in the biochemical composition and blood cells, leading to disruptions in the work of antioxidant defense systems.

Experiments on animals have proved the close relationship between the content of CIPE and their impact on the body and the need for regulation and rationing of this indicator in finished products at the stages of their circulation and realization to ensure the safety of fast food products.

REFERENCES

Abdel-Aal, M. H., & Karara, H. A. (1986). Changes in corn oil during deep-fat frying of foods. *Lebensmittel-Wissenschaft + Technologie*, *19*, 323–327.

Andrikopoulos, N. K., Dedousis, G. V. Z., Tzamtzis, V., Chiou, A., & Boskou, G. (2002). Evaluation of medium polarity materials isolated from fried edible oils by RP-HPLC. *European Journal of Lipid Science and Technology*, *104*(2), 110–115. doi:10.1002/1438-9312(200202)104:2<110::AID-EJLT110>3.0.CO;2-C

Guide for the Care and Use of Laboratory Animals. (2011). National Research Council, National Academy Press.

Kravchuk, N. N. (1981). *Dynamics of the fractional composition of the "Frying" culinary fat and sunflower oil when deep-fried product. In Coll. scientific. tr.* KTEI.

Loskutova, Z. F. (1980). Vivarium. *Medicine*, 93.

Markaverich, B. M., Crowley, J. R., Alejandro, M. A., Shoulars, K., Casajuna, N., Mani, S., Reyna, A., & Sharp, J. (2005). Leukotoxin Diols from Ground Corncob Bedding Disrupt Estrous Cyclicity in Rats and Stimulate MCF-7 Breast Cancer Cell Proliferation. *Environmental Health Perspectives*, *113*(12), 1698–1704. doi:10.1289/ehp.8231

Marmesat, S., Mancha, M., Ruiz-Méndez, M. V., & Dobarganes, M. C. (2005). Performance of sunflower oil with high levels of oleic and palmitic acids during industrial frying of almonds, peanuts and sunflower seeds. *Journal of the American Oil Chemists' Society*, *82*(7), 505–510. doi:10.100711746-005-1101-6

Marquez–Ruiz, G. (2006). Nutritional and physiological effects ofused frying oils and fats. In Deep frying: Chemistry nutrition and practical applications (2nd ed.). Champaign, IL: American Oil Chemists' Society.

Nosova, A. S. (2013). *Improving the quality of fat for the production of frying products* (PhD dissertation). State University – UNPK.

Recommendations for euthanasia of experimental animals: Part 1, Part 2. (1996). *Laboratory Animals, 30*, 293-316.

Rimac-Brncic, S., Lelas, V., Rade, D., & Simundic, D. (2004). Decreasing of oil absorption in potato strips during deep fat frying. *Journal of Food Engineering*, *64*(2), 237–241. doi:10.1016/j.jfoodeng.2003.10.006

Shilman, L. Z. (1966). *Study of the comparative resistance to oxidation of some vegetable oils under hydrothermal action* (PhD dissertation).

Skopichev, V. G. (2003). *Animal physiology and etiology.* Kolos.

Terzieva, V. (1990). Study of the quality of fat when frying vegetable and meat semi-finished products. *Inst. Stores. Taste. prom., 37*(3), 105–113.

Tsynko, T. F. (2005). Diagnosis of diseases by blood and urine tests. Phoenix.

Velasco, J. (2005). Quantitation of short-chain glycerol-bound compounds in thermoxidized and used frying oils. A monitoring study during thermoxidation of olive and sunfloweroils. *Journal of Agricultural and Food Chemistry*, *53*, 4006–4011. doi:10.1021/jf050050t PMID:15884831

Velasco, J., Marmesat, S., Bordeaux, O., Márquez-Ruiz, G., & Dobarganes, C. (2004). Formation and evolution of monoepoxy fatty acids in thermooxidized olive and sunflower oils and quantitation in used frying oils from restaurants and fried-food outlets. *Journal of Agricultural and Food Chemistry*, *52*(14), 4438–4443. doi:10.1021/jf030753f PMID:15237949

Chapter 8
Research of the Safety Indicators of Some Instant Concentrates

ABSTRACT

In this chapter, the range of food concentrate products is currently quite diverse; the popularity of these products is great. In this regard, the question arises: Can food concentrates constitute a worthy substitute for a healthy diet? How safe is regular consumption of these products for human health?

INTRODUCTION OF THE CHAPTER

The materials in this chapter are based entirely on the authors' own research.

Most food concentrates are safe from the point of view of microbiological parameters, for example, Wójcik-Stopczyńska (Wójcik-Stopczyńskaet al., 2002) studied a line of instant soup concentrates, the results of which showed that the microbiological quality of instant soup concentrates was in accordance with the standards.

The microbiological safety of this group of products is due to the low moisture content. The amount of moisture in food concentrates does not exceed 10-12% (in some of them, for example, in corn flakes, it is below 5%), and therefore they are not suitable material for the development of microorganisms.

DOI: 10.4018/978-1-7998-8197-1.ch008

Several basic technological processes are used in the food concentrates production:

- mechanical processing;
- heat treatment, including blanching in fat;
- drying.

According to physical and chemical parameters, food concentrates of soups and hot second course should satisfy the following requirements: humidity of vegetable concentrates is not more than 12%, legumes, cereals and noddle products - 10%, egg - 8.5%.

Other important issues are changes in the components of concentrates in the process of their preparation.

Intensive mechanical and thermal effects on raw materials at their processing in the production of food concentrates leads to the fact that the nutrients in them are sufficiently freed from fiber, the cell walls of the raw materials are severely destroyed, starch is klisterized and dextrinated, proteins are denatured.

Due to the effects of high temperature and water, nutrients (mainly proteins and carbohydrates) in concentrates are partially hydrolyzed. And although it determines their best intake by the body, there is a likelihood of losing biological value and effectiveness due to harsh technological regimes.

Dehydration of raw materials is one of the processes that form the quality of food concentrates.

By the method of supplying heat to raw materials, the following types of artificial drying are distinguished:

- convective - by direct contact of the product with a drying agent, most often air;
- contact - heat transfer from the heat carrier to the product through the wall separating them;
- radiation - heat transfer by infrared rays;
- dielectric - by currents of high and ultrahigh frequency;
- vacuum and its variety - sublimation (Ginzburg A.S., 1973; Atanazevich, V.I., 2000).

The most common and simplest type of drying is convection.

High-tech process is as freeze-drying (sublimation), which allows one to almost completely (up to 95%) preserve nutrients, vitamins, trace elements, the original form, natural smell, taste and color. This is one of the most important advantages of sublimation. The method allows one to avoid destruction of the product structure, to restore the freeze-dried products, as they have a porous structure.

Thermal processes of food concentrates production cause the inactivation of enzymes, as a result of which enzymatic changes in food concentrates occur extremely slowly and in many cases are completely suspended. The drying technology is often associated with deep chemical changes and the loss of valuable natural components of vegetables.As a result of the active action of hydrolytic enzymes and chemical catalysts (for example, hydrogen ions), the ratio between simple and complex carbohydrates changes during drying. Due to oxidation, di- and tricarboxylic acids are produced from sugars, and when they are decarboxylated, monobasic acids are produced.

When heated, a complex set of reactions takes place; some of the sugars are caramelized. Changes in the protein complex and pectin substances affect the regenerative ability of dried vegetables.

During drying, various volatile substances (esters, alcohols, aldehydes) are partially removed, which also leads to a decrease in product quality.

In addition, new volatile products are formed and removed as a result of biochemical reactions and chemical destruction processes. Due to deamination, some amino acids can be converted to ammonia and volatile aldehydes.

At the same time, with the loss of moisture, a partial increase in dry matter is observed due to oxidative reactions with the addition of oxygen; therefore, the total weight loss during drying is partially compensated.

Biologically active substances, such as ascorbic, folic acid, carotenoids, and some polyphenols are subjected to more significant decomposition during drying. Ascorbic acid undergoes the greatest destruction.

The problem of the persistence of aromatic components of products, many of which are very volatile and therefore can be lost in the process of dehydration, is of great interest.

The organoleptic characteristics of many dried vegetables during prolonged storage practically do not change in the presence of air.

Due to the difference in the structure, physicochemical and biochemical properties of the drying products, the nature of the change in their quality indicators during storage, has individual characteristics specific to a particular product (Burich O., Berki F.,1978; Sacilik K. et al., 2006).

During long-term storage of dried fruits and vegetables, even under optimal conditions, a decrease in their quality due to oxidative damage is observed. The oxidation rate increases with increasing humidity of the product. Dried vegetables contain a small amount of fat. The rancidity of fat in vegetables is due to oxidation by atmospheric oxygen.

It should be noted that almost all concentrates, with the exception of sweet food concentrates, contain a fatty component added to the product or absorbed by it as a result of blanching in fat. The issue of oxidation of the fat component in foods with an extended shelf life is not well understood. The relevance of studies of the fat component safety is due to the need and feasibility of ensuring the high quality of instant concentrates and their stability during storage.

At present, when the attention of scientists is attracted to the problem of fast food consumption and its effect on the body, there is a problem of the excessive amount of salt in these products and the possibility of modifying them. A large number of scientific papers are devoted to the study of salt content in fast food products (Menyanu E. et al., 2019). According to some scientists, reducing salt intake is a cost-effective public health measure to reduce the global burden of noncommunicable diseases (Farrand C. et al., 2017). This is confirmed by other studies. J. Park et al. revealed that consumption of instant noodles can lead to increased salt and caloric intake, mainly due to fats(Park J. et al., 2011).

Based on the foregoing, the study of some safety indicators of instant concentrates and their likely impact on the human body with regular consumption is relevant.

The purpose of the study is to substantiate the issues of safe consumption of food concentrates.

The tasks set by the authors included:

- determination of safety indicators for the fat component of soup concentrates and hot second courses, the most popular in the world;
- determination of the functional and technological properties of starch polysaccharides of kissel concentrates from natural raw materials that do not require heat treatment and their comparative evaluation with analogue ready meal;
- study of the effect on the morphological and functional parameters of the organism of experimental animals with regular consumption of concentrates of kissel of industrial production in comparison with kissel cooked using traditional technology.

The objects of the study: the most popular in the Russian Federation fast food products of industrial production, produced using various processing methods, including those with a large proportion of the fat component, various price categories, various manufacturers:

- instant soups. Samples: IN (production Croatia), IT (production / owner Russia / Switzerland), IK (production / owner Russia / Switzerland), IX (production / owner Russia / Spain), IL (production / owner Russia / Netherlands); IW (production / owner Russia / Switzerland), IV (production / owner Russia / Spain), IM (production Russia);
- instant potatoes. Samples: PI and PII (production / owner Russia / International group of companies).
- instant noodles;
- concentrates of sweet dishes (kissel).

Methods Used to Study the Fat and Fat Component of Fast Food Products

In the first case the subject of the study was the physicochemical changes in the fat component of fast food products. The objects of the study were the most popular fast food products in the Russian Federation - concentrates of soups, mashed potatoes, and instant noodles of popular brands.

The shelf life of the concentrates is 12 months. The opening of sealed packages purchased through retail chains was carried out in the initial period of the shelf life - in the second or third month of storage.

Fat for analysis was isolated from the product by the extraction-weight method (GOST 31902-2012 'Methods of determination of fat weight fraction'.).

In the extracted fat, the fatty acid composition was determined by gas-liquid chromatography of methyl esters of fatty acids (GOST R 51483-99 'Vegetable oils and animal fats. Determination by gas chromatography of the mass fraction of methyl esters of individual fatty acids to their sum'). Fatty acid methyl esters were prepared according to (GOST R 52110-2003 'Vegetable oils. Methods for determining the acid number') and the averaged fatty acid composition of the studied products was analyzed.

The normative documentation of the Customs Union (TR CU 0211/2011, TR CU 033/2013) regulates only the determination of the peroxide number, which characterizes the accumulation of primary products of fat oxidation. The peroxide number of the fat component of the product was determined by the iodometric method (GOST R 51487-99 'Vegetable oils and animal fats. Method for determination of peroxide value').

To assess the degree of hydrolysis and oxidation of the fat component, the acid number of the released fat was determined by the titrimetric method (GOST 31933-2012'Vegetable oils. Methods for determination of acid value'), the content of secondary oxidation products insoluble in petroleum ether was determined according to (Rzhehin V.P., Sergeev A.G., 1967), and the content of epoxides by reaction with concentrated phosphoric acid was determined according to (Stopskij V.S.et al., 1986) All studies were carried out immediately after opening the packages with concentrates.

Determination of acid number is important for determining the degree of fat hydrolysis. This indicator is of particular importance, since the process of noodle dehydration during frying can be accompanied by intensive hydrolysis of fats, which provokes the accumulation of di- and monoglycerides, which, according to modern data, can be precursors of the dangerous toxicant 3-chloropropanediol (3-MCPD) (Jedrkiewicz R. et al., 2016).

Statistical processing of the described studies complies with international requirements and is carried out in accordance with ISO 5725 (parts 1-6) and ISO / IEC 17025 - 99. All test results are processed by standardized measurement methods and meet the requirements of trueness and precision of measurements. The replication of measurements is at least 5 times, a confidence probability of the accuracy of test results is 95%.

The work was carried out on the basis of the research laboratories of the Federal State Budget Educational Institution of Higher Education "Saratov State Agrarian University named after N.I. Vavilov", Saratov.

MAIN FOCUS OF THE CHAPTER

8.1 Instant Soups

This chapter discusses the issues of safe replacement of traditional dishes with concentrates and their possible impact on the human body with regular consumption.

Instant soups.Instant soups are produced according to two main technologies - by dehydrogenation and by sublimation. The first technology is of the greatest distribution, as it is less expensive. The second technology is more expensive.

Most of the instant soups on the Russian market are produced by dehydration (up to 90%) and are a mechanical mixture of the main component with fat, a small amount of meat and various flavors.

The composition and components included in the recipe for instant soups, indicated on the package are regarded. Table 8.1 presents the ingredient compositions of soup concentrates.

Analysis of the components included in the concentrates indicated on the package showed that in addition to the main component (noodle products, cereals, legumes), almost all instant soups contain spices or their extracts, as well as dried vegetables, such as carrots and onions. It should be noted that dried onions are quite stable, while there is a possibility of loss of such minor components as phytoncids and some of the components responsible for aroma. In carrots with any method of drying (dehydration) beta-carotenes are unstable, which leads to discoloration of carrots and oxidative damage, which manifests itself in the presence of a rancid odor. Most producers use carrot blanching or boiling. It practically prevents oxidative reactions, but reduces the nutritional value and the ability to digest when quickly cooking.

Soup concentrates contain vegetable fats, palm oil or its fractions, in some cases combined fats (a mixture of vegetable and animal fats).

Special attention should be paid to excipients of the ingredient composition of concentrates. Although it is difficult to call them accessory, because, this is the main "trump card" of instant concentrates, which forms the organoleptic properties of concentrates.

They all contain flavoring and aromatic additives, such as glutamate, sometimes in conjunction with synergists and flavorants. According to experimental data of K. Ackroff, A. Sclafani, 2016(Ackroff K, Sclafani A.,2016), glutamate enhances consumer preferences by forming an impression of taste. The authors obtained such data during simulated experiment, in vivo.

In addition, complex food additives are used, for example, flavorant Chicken, which includes maltodextrin, glucose, flavor and aroma enhancer (E621 - sodium glutamate, E627 - sodium guanilate, E631 - sodium inosinate), soy protein hydrolyzate, salt, powdered methileaf,andflavorant (Chicken)).

Sodium guanylate is a powerful modifier of taste and aroma, which significantly increases the organoleptic characteristics of the product. In comparison with other additives of a similar action, disodium guanylate is much more effective. In this case, the additive acts at the cellular level, taking part in metabolic processes. That is why its consumption should be strictly dosed. Disodium guanylate is a preservative, its introduction into the composition increases the shelf life of the finished product. It is the effect of sodium guanylate on metabolic processes in the body that explains the presence of negative effects from its consumption in food: when E627 is broken down, purines are formed in the body - substances that cause the gout progression.Frequent consumption of a taste modifier can cause the development of hypertension, skin allergic reactions, and even Quincke's edema. It affects the work of the gastrointestinal tract, causing indigestion, flatulence, diarrhea, and cramping. Other side effects include headaches, insomnia, apathy, dehydration, and loss or uncontrolled increase in appetite.

Sodium inosinate is one of the strongest flavor enhancers causing persistent addiction. Sodium inosinate is typically used in combination with sodium glutamate. This food supplement has not been studied and there is no direct ban on it in most countries. In the United States, the maximum daily dose of sodium inosinate is 5 g. The consumption of a large amount of the additive can cause some negative consequences and problems with the gastrointestinal tract.

Yeast or its extracts can be considered as the safest flavor enhancers.

Thickeners are always present in the formulation of concentrates. They form the structural properties of the final product and, accordingly, participate in the taste formation. In the ingredient composition, modified starches, sometimes others, for example, xanthan gum, are used as thickeners.

There are up to 72 product categories that use modified starch. Unlike native plant starches, modified starches are food additives. This group of food additives includes products of fractionation, destruction and various modifications of native plant starches, which are mainly a mixture of two fractions of homoglucans (glucose polymers) of a linear and branched structure. The chemical structure and properties of these fractions, as well as their ratio in native starch, depending on the type of plant source (potatoes, rice, corn, etc.), determine the main technological property that native starch exhibits in food systems. Modified food starches are relatively safe additives, the dosages of which are regulated by technological considerations.

Thus, it should be noted that soup concentrates have different ingredient compositions of excipients and additives. An effective approach to assessing the safety of natural flavoring complexes (NFC) was published in 2005 by the Expert Group of the Flavor and Extract Manufacturers Association (FEMA), which provides for a reassessment of the safety of flavoring and aromatic

Table 1. Composition of soups according to the traditional technology and concentrate soups (Personal analytical data)

Sample	Production	Composition	
		Mainingredient	Excipients
Chickenbrothvermicellisoup	Traditionaltechnology	Chicken, vegetables, vermicelli, greens, chicken broth	Salt
IN	Croatia	Vermicelli, iodized salt, palm oil, dried vegetables (carrots, onions, celery), dried chicken extract, chicken fat	Taste enhancers (monosodium glutamate, sodium inosinate), corn starch, sugar, dye (caramel color I), parsley leaf
IT	Russia /Switzerland	Pasta, dried vegetables (carrots, potato flakes, onions, parsnips, garlic, paprika, celery, tomatoes), dried greens, sunflower oil	Iodized salt, sugar, spices, caramelized burnt powdered sugar, powdered beef
IK	Russia /Switzerland	Noodle products, dried vegetables and greens	Salt, food flavor "Chiken", sugar, vegetable oil, yeast extracts, turmeric, citric acid
IX	Россия/Spain	Croutons (wheat bread, palm fat), carrot, parsley, chicken meat, onion	Thickener - modified potato starch, vegetable powdered fat, salt, yeast extracts, sugar, sunflower oil, dyes (caramel color I, riboflavin), bay leaf, celery, nutmeg
IL	Russia /Netherlands	Wheat croutons, dried greens and vegetables	Potato starch, wheat flour, salt, sugar, palm oil, flavoring, spices, lactose, sodium caseinate, sodium citrates, dried boiled chicken
Pea cream soup	Traditionaltechnology	Shelled peas, carrot, onion, vegetable oil, chicken / beef broth	Salt, blackpepper, bayleaf
IW	Russia /Switzerland	Pea flour, wheat flour, dried vegetables	Palm oil, salt, yeast, maltodextrin, iodized salt, sugar, flavors (grill, yellow peas, culinary base, beef, boiled onions, asparagus, peas) and natural (black pepper extract), spices and flavoringherbs, glutamate, guanilate and sodium inosinate, xanthan gum, citric acid, sodium citrates

Continued on following page

Table 1. Continued

Sample	Production	Composition	
		Mainingredient	Excipients
IV	Russia / Spain	Croutons (wheat bread, palm fat), carrot, onion, peas, garlic, parsley	Thickener - acetylated starch, powdered vegetable fat (glucose syrup, palm fat, milk protein), salt, yeast extracts, sugar, sunflower oil, dyes (paprika extract, riboflavin), bay leaf, flavorings (celery, soy), clove.
IM	Russia	Boiled and dried peas, dried potato flakes, dehydrated pork, dried carrots, dried onions	Sunflower oil, flavorant Bacon, sugar, dried parsley, dried garlic, bay leaf, black pepper extract

additives. FEMA experts proposed an updated procedure, emphasizing more stringent requirements for those components that have not yet been adequately studied taking into account their genotoxic potential (Cohen S.M.et al., 2018).

Nevertheless, a modern person eats a large number of such substances, which cumulative effect on the body is not fully understood. One cannot but take into account the fact that concentrates consumption can change the taste preferences of a person.

Table 2 presents the ingredient compositions of soups according to the traditional technology and concentrate soups in order to compare their nutritional and energy value.

Table 2 shows that the nutritional value of the finished dish weighing 100 g is significantly lower than that of dishes according to traditional technology. Almost all of the studied groups of concentrates, which are analogous to chicken soup with vermicelli, have a significant deviation in the protein component to a lesser extent from the difference of 2.16 and 2.55 times for a sample of IN and IK groups. A similar situation is observed with concentrates, which are analogous to pea soup. Here the difference is even more significant, as bean soups are a source of protein close to complete, and the protein content in concentrate soups is 10 times less.

This suggests that concentrates cannot completely make up for the protein dietary requirements expected for such dishes as soups.

Significant fluctuations are observed in the fat content per 100 g of ready-to-eat soup, both in a smaller direction. This is typical for pea concentrate

Table 2. Nutritional and energy value of traditional and concentrate soups (Personal analytical data)

Sample	Manufacturer	Nutritionalandenergyvalue	
		Per 100 g of dry product	100 gof ready soup
Chickenbrothvermicellisoup	Traditional technology	-	protein – 2.81 fat – 2.01 carbohydrate – 5.08 45,82 kcal
IN	Croatia	protein – 13. fat – 10. carbohydrate – 52 360 kcal	protein – 1.3. fat – 1.0 carbohydrate – 5.2 36 kcal
IT	Russia/Switzerland	protein – 8.4 fat – 1.5 carbohydrate – 62 311 kcal	protein – 0.44 fat – 0.08 carbohydrate – 3.36 16.8 kcal
IK	Russia /Switzerland	protein – 14.5 fat – 2.5 carbohydrate – 62 330 kcal	protein – 1.1 fat – 0.17 carbohydrate – 4.34 23.12 kcal
IX	Russia / Spain	protein – 4 fat – 17 carbohydrate – 57 410 kcal	Б – 0.31 fat – 1.45 carbohydrate – 5.1 69.7kcal
IL	Russia /Netherlands	protein – 7 fat – 17 carbohydrate – 55 400 kcal	protein– 0.62 fat – 1.51 carbohydrate – 4.88 35.5 kcal
Pea cream soup	Traditional technology	-	protein – 10.3 fat – 1.3 carbohydrate – 20.4 ПВ- 5 Зола – 2.4 134.5 kcal
IW	Russia /Switzerland	protein – 5.4 fat – 11 carbohydrate – 51 356 kcal	protein– 0.57 fat – 1.16 carbohydrate – 5.38 37.58 kcal
IV	Russia /Spain	protein – 4 fat – 18 carbohydrate – 59 410 kcal	protein – 0.68 fat – 3.06 carbohydrate – 10.03 69.7 kcal
IM	Russia	protein – 14.0 fat – 8.0 carbohydrate – 58.0 360 kcal	protein– 0.98 fat – 0.56 carbohydrate – 4.06 25.2 kcal

soups. A fat content in group IV soups is 2.35 times higher than in traditional soups, while in IM group it 2.31 times less.

Soups in IT-group contain 25 times less fat compared to traditional chicken noodle soup. The smallest difference is in the sample of the IL-group, the amount of fat per 100 g is 1.33 times less than that of the soup according to the traditional technology.

Concentrates of soups similar to chicken noodles with vermicelli have no significant deviations in carbohydrate content. And in concentrate soups with legumes, there is a significant difference in the direction of reduction.

Table 2 shows that, in terms of nutritional and energy value, concentrate soups cannot be a full-rate analogue to soups according to the traditional technology, and therefore do not make up for the need for nutrients. These concentrates are only able to satisfy temporary hunger, creating the illusion of temporary satiety due to stabilizing additives. This is especially true for instant soups.

8.1.1 Study of The Content of Fats in Instant Soups

Figure 1 shows the fat content of soup concentrates in accordance with the data indicated on the package.

Figure 1. The mass fraction of the fat component indicated by the manufacturer on the package per 100 g of dry product. (Personal research data)

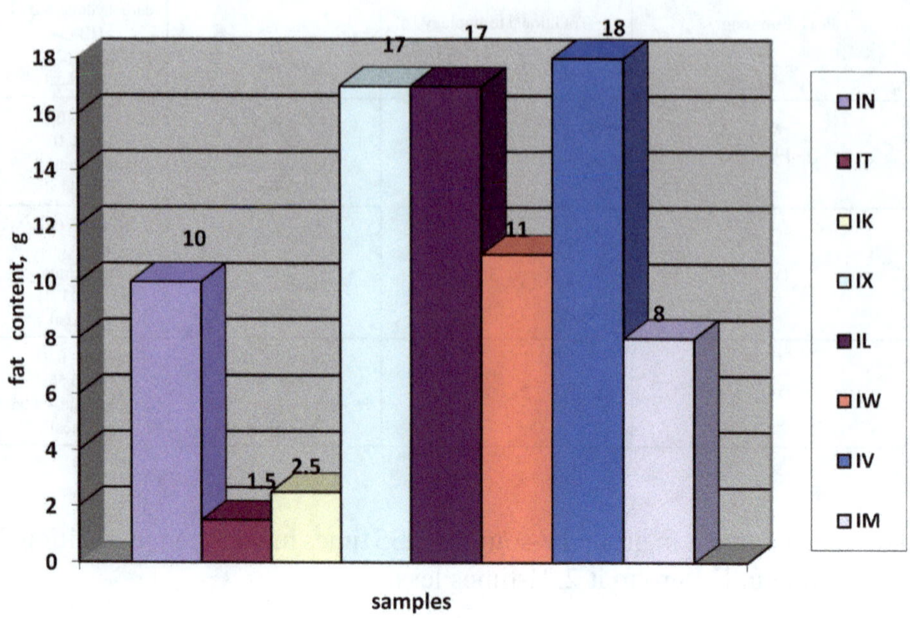

According to Table 2 and Figure 1, soups differ in fat content. The mass fraction of the fat component ranges from 1.5 to 18 g per 100 g of dry product. On average, the fat content is consistent with traditional recipes.

Table 3 summarizes the fatty acid composition of instant soups. Analyzing the data obtained by determining the mass fraction of methyl esters of individual fatty acids to their sum by gas chromatography, with the data indicated on the packaging (Table 1), their incomparability should be noted. For example, according to the fatty acid composition, the IN-sample contains mainly palm oil.

At the same time, the manufacturer indicated chicken fat in the composition. According to the fatty acid composition, IM-sample is close to sunflower oil, as declared by the manufacturer. In IX- and IV-samples of instant soups, there is a mixture of various fats with a predominance of palm oil. It should be noted that soup concentrates do not differ in a balanced fatty acid composition.

Table 3. The fatty acid composition of some instant soups (Personal research data)

Sample	Thefattyacidcomposition										
	Caprylic C8:0	Capric C10:0	Lauric C 12:0	Myristic C14:0	Palmitic C16:0	Palmitoleic C16:1	Stearic C18:0	Oleic C 18:1	Linoleic C18:2	Linolenic C18:3	Behemic C 22:0
IN				1.6	49.3	0.6	4.7	34.2	9.6		
IM					9.8		4.1	25.1	60.3		0.7
IV	0.9	0.7	5.2	2.7	31.8		4.4	33.3	20.7	0.4	
IX	0.7	0.6	4.2	2.3	31.8		4.6	33.1	22.4	0.3	

The results of studies of safety indicators for fats contained in instant soups are of particular interest (Table 4). To determine the safety of the fat component of soup concentrates, it is possible to study two regulated indicators: the content of free fatty acids and of peroxides. The mass fraction of oxidation products insoluble in petroleum ether is subject to control only in deep frying fats, however, it can and should serve as a criterion for assessing the degree of fat changes in the technological process and during storage of products.

It should be noted that the data of peroxides content, the indicator that is controlled by the legislation of the Russian Federation are within the normal range in all soup concentrates, with the exception of IV-sample, where it is close to the upper limit of the norm.The acid number indicates the course of the hydrolytic decay of fat into glycerin and free fatty acids. The consequence of this decay is not only the loss of biological effectiveness, but also the appearance of an unpleasant greasy taste and smell. It should be noted that fat component of all soup concentrates undergoes hydrolytic changes with varying degrees of intensity.

As for the content of oxidation products insoluble in petroleum ether, in all samples of soup concentrates, with the exception of IN-sample, this indicator is equal to, approximate, or exceeds the norm at which deep frying fats that underwent repeated heat treatment are discontinued.

Table 4. Safety indicators of the fat fraction included in instant soups (Personal research data)

Sample	Acid number, mgKOH/g	Peroxide number, meq½O/kg	Mass fraction of oxidation products insoluble in petroleum ether, %	Epoxy content, meq/kg
IN	0.8	1.1	0.4	28.6
IM	5.6	2.1	1.3	19.3
IV	1.0	7.9	1.0	36.1
IX	1.1	6.6	0.9	28.2

Almost all samples, with the exception of IM-sample, contain a great number of epoxides.

The accumulation of concentrates of thermostable secondary oxidation products in the fat component is characteristic of the fat oxidation during heat treatment. The production of highly polar compounds insoluble in petroleum ether and epoxides was discovered during the study of deep fried dough products, as well as during the storage of confectionery products with a high fat content - Kurabye cookies, shortbread cookies, national dough products fried in deep frying fat and others. In a biological experiment on animals, in our previous studies, it was shown that fats containing more than 1% of secondary oxidation products, when systematically consumed, have a negative effect on the organs of the gastrointestinal tract, sharply reduce the level of red and white blood cells, cause the accumulation of cholesterol and bilirubin (Simakova I.V.,2015). The possibility of the production of such compounds in soup concentrates is a serious concern.

Soups are a symbol of well-being and health. Preserving the culture of soup consumption is an important aspect in the formation of a food culture in general. Obtained data indicate the need for a critical assessment of the technology for the production, packaging and storage of soup concentrates, as well as the feasibility of amending the Codex Alimentarius set of regulatory international standards with the aim of additionally monitoring the safety of the fat component of soup concentrates and changing the shelf life of these products during storage.

8.2 Instant Potato Puree

8.2.1 Technology of Production of Instant Potato Puree

During the heat treatment of potatoes, as well as during its storage before processing, biochemical reactions occur in tubers under the action of the oxidation-reduction enzyme amylase and phosphorylating esterases.The main directions of changes in carbohydrates are the formation of sugars and starch. The enzymes involved in these reactions, can also catalyze the reverse process - the synthesis of more complex carbohydrates (up to starch and hemicelluloses) from less complex ones.In boiled potatoes, there is no increase in sugar content during further technological processing, which indicates the inactivation of enzymes involved in carbohydrate metabolism.

The most heat-resistant enzyme is peroxidase. For its inactivation, plant tissue must be heated to a temperature beyond 75 °C. The completeness of inactivation also depends on the duration of the heat exposure. A qualitative reaction to peroxidase activity (by the action of guaiacol in the presence of hydrogen peroxide) can indirectly serve as an indicator of inactivation of all other enzymes, therefore, the reaction to peroxidase is a method of controlling the process of potatoes blanching and boiling.

Along with the physicochemical changes of potatoes during blanching and boiling, which favorably affect the quality of dried foods and their shelf life, undesirable changes also occur: loss of nutrients, especially when blanching in water, partial destruction of vitamins, mainly of vitamin C, partial destruction of potato cells.

Preservation of whole potato cells is one of the conditions for the production of dry floury potato puree.

It is known that the methods and modes of boiling influences on the preservation of cells and the nutritional value of potatoes.

Depending on the type of coolant, steam and water boiling are distinguished, depending on the multiplicity - single and double with intermediate cooling. Many researchers recommend the use of double boiling potatoes in technological schemes for the production of dry potato puree in order to strengthen potato cells and prevent their destruction at further stages of the technological process.

When thermal influence of the middle plates of the potato tissue on the protopectin, a process of hydrolysis is marked. In this case, water-insoluble protopectin passes into a simpler form - water-soluble pectin. The middle plate dissolves, and the cells are disconnected.

However, under the long-term exposure of hot water or steam, which is necessary for the complete dissolution of the middle plates over the entire thickness of the tuber, the destruction of the cell membrane due to the difference in the osmotic pressure inside the cell and in the intercellular space during heating can occur. Therefore, to prevent cell destruction during heat treatment, intermediate cooling of potatoes is carried out before cooking.

Changes in starch resulted from blanching stabilize during cooling. Cooling causes the retrogradation of starch amylose. Retrograded amylose becomes less soluble at low temperatures, and swollen starch becomes denser, which conduces cells protection from mechanical damage.

Lower cooling water temperatures and longer cooling times have a positive effect on the consistency of potato puree. The cooling time is 20-40 minutes.

Stabilization of starch properties resulted from blanching and cooling, is maintained at the subsequent stages of the process, positively affecting the quality of potato puree.

In short-cycled technological schemes for the production of dry potato puree, when the potato is subjected to insignificant mechanical and thermal exposures, it is recommended to use a single potato boiling.

Comparative studies of potato boiling showed that a single boiling preserves the biologically active substances of the potato to a greater degree and determines a higher nutritional value of the product.

However, in the vast majority of countries, in order to produce floury potato puree, double boiling with intermediate cooling is used. Blanching is carried out in water, boiling - by saturated steam without excessive pressure, which is fed into the cooking apparatus through nozzles or bubblers.

The most important role in the production of dry potato puree is played by drying.

Potato has thermolabile properties, so drying should be carried out under conditions that prevent the occurrence of irreversible processes leading to a deterioration in the product quality. In the production of dry potato puree, changes in the color are unacceptable that may occur during the interaction of amino acids and sugars; the drying agent must have a temperature at which caramelization and carbonization of the product surface would not occur.It is also necessary to ensure minimal destruction of potato cells and preserve the main potato nutrients: vitamins, proteins, carbohydrates. The natural organoleptic properties should not be changed.

The main indicator of the quality of dry potato puree is the consistency of the reconstituted potato puree produced from it, which depends mainly on the content of free starch in it. The high content of free starch, due to the destruction of the cell walls of the potato tissue during the processing treatment, gives a paste-like consistency to the reconstituted puree.

It was found out that when the content of free starch is less than 5% (for absolutely dry matter), the consistency of reconstituted puree is characterized by low viscosity, from 5 to 10% - by medium viscosity, and from 10 to 15% - by high viscosity as well as by unsatisfactory organoleptic characteristics. Depending on the shape and size of the particles, dry potato puree is produced in the form of grains, flakes and granules.

Potato grain is a fine-grained product with a moisture content of up to 12% with a grain size of up to 1 mm, white or light cream in color. It is quickly reconstituted to potato puree during cooking. Water with a temperature of 80-85 °C is added to the dry semi-finished product (4-5 parts of water are added to 1 mass part of the product) and the mixture is let sit for 2-3 minutes.

The volumetric mass of potato grains is 0.5-0.7 kg / l, that is 2.5 times more than the mass of potato flakes; so when packing, it requires less packaging and reduced transportation costs.

Potato flakes are very thin (0.2-0.3 mm) petals, white or light cream in color, no larger than 10 mm. The moisture content of the finished product, packaged in sealed containers, is not more than 8%, in unpressurized containers - not more than 12%.

The volumetric mass of the product is 0.2-0.3 kg / l. The dry product is instantly reconstituted in potato puree when pouring with hot water.

Potato granules are cylinders with a diameter of 1-3 mm and a length of 5-25 mm, white or cream in various shades, characteristic of the corresponding varieties of potatoes. The moisture content of the finished product, packaged in sealed containers, is not more than 8%, in unpressurized containers - not more than 12%. Their bulk mass is 0.3-0.4 kg / l.

To cook potato puree, the product is boiled or kept in thermostatic conditions (pour hot water) for 10-15 minutes.

The United States organized an industrial production of a product, which is a type of grains, called "flakelets" according to the following scheme.

After washing, the potatoes are subjected to alkaline cleaning, sulfitation, inspection, cutting into mugs 12-15 mm thick, water blanching at a temperature of 75-80 ° C for 20-30 minutes, cooling with cold water, steam cooking until it is cooked and mashed.

At mashing, antioxidants and emulsifiers are introduced into the product. The resulting potato puree is divided into two parts. One part of it is subjected to conductive drying to produce dry flakes, and the second one is mixed with dry potato flakes; the mixture is cooled and granulated into grains up to 2 mm in size. The granular product is subjected to convective drying to a moisture content of 5-7%.

Dry potato puree in the form of flakes is produced in many foreign countries.

The consistency of reconstituted puree produced from it depends mainly on the content of free starch in it.

The composition, nutritional and energy value of potatoes cooked according to the traditional recipe and potato puree concentrate are presented in Tables 5 and 6.

Table 5. Composition of potato puree cooked according to the traditional recipe and instant potato puree (Personal analytical data)

Sample	Manufacturer / accessory	Composition	
		Mainingredient	Excipients
the traditional recipe	According to the traditional recipe	Boiled potatoes, mashed, milk, butter	Salt
PI	Russia / International group of companies	Potato puree: potato flakes, vegetable-based dry cream substitute (vegetable fat, glucose syrup, milk protein, emulsifier E471), Braised chicken: chicken, drinking water, onions, carrots, salt, dried garlic, light wheat flour, sunflower oil, sugar, black pepper, turmeric	Salt, sugar, fried onions, dried carrots, scallions, and parsley, flavor and aroma enhancers (glutamate, guanylate, sodium inosinate), flavourant "Chicken", antioxidant (E223)
PII	Russia / International group of companies	Potato flakes (potato (99%), vegetable-based dry cream substitute (vegetable fat, glucose syrup, milk protein, emulsifier E 471), wheat croutons (premium wheat flour, vegetable oil, salt, yeast, antioxidant - rosemary extract, preservative (E282)	Fried dried onions (onions, vegetable oil, premium wheat flour, salt), sugar, dried parsley, flavoring, turmeric

Table 6. Nutritional and energy value of potato puree cooked according to the traditional recipe and instant potato puree (Personal analytical data)

Sample	Manufacturer	Nutritionalandenergyvalue	
		Per 100 g of dry product	100 g of ready potato puree
Potato puree	According to the traditional recipe	-	Protein – 2. 1 Fat – 0.8 Carbohydrate – 14.7 75.00 kcal
PI	Russia / International group of companies	Protein – 8.5 Fat – 13.0 Carbohydrate – 66.0 420 kcal	Protein – 3.86 Fat – 5.91 Carbohydrate –30 191 kcal
PII	Russia / International group of companies	Protein – 7.4 Fat – 4.2 Carbohydrate – 70.5 349 kcal	Protein – 3.36 Fat – 1.91 Carbohydrate –32.05 159 kcal

It is obvious that, as in the case with soup concentrates, a significant amount of excipients is used to create taste and aroma in potato puree concentrates: spices, dried vegetables, flavorings, emulsifiers, flavor and aroma enhancers, etc. The manufacturer also indicates the composition of cream substitute in which vegetable fats predominate.

8.2.2 The Study of Safety Indicators of The Fat Component of Potato Puree Concentrates

Based on the analysis of nutritional value, it should be noted that puree concentrates differ significantly from the traditional recipe, especially in terms of the fat and carbohydrate component, the difference in protein occurs due to the excipients, crackers, and poultry.

The energy value of these products is 2.1 - 2.5 times higher than the traditional recipe.

Figure 2 and Table 8 presents research data on the quantitative and qualitative fat component of potato puree concentrates.

According to Figure 2, the content of the mass fraction of extracted fat in the test samples exceeds the declared on the package.

The fatty acid composition of potato puree concentrates is presented in Table 7.

Figure 2. Mass fraction of fat in puree concentrates, %(Personal research data)

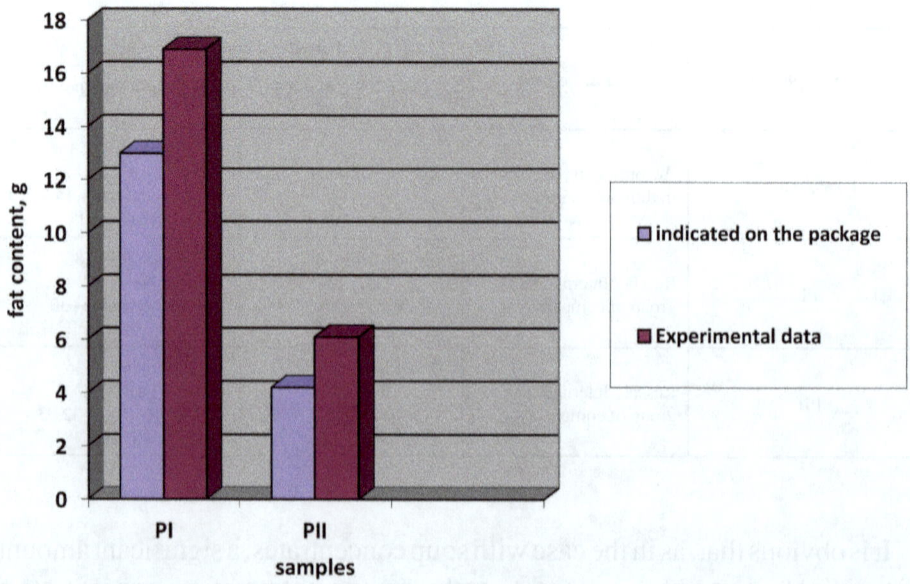

Table 7. The fatty acid composition of fat component in instant potato puree (Personal research data)

Sample	Thefattyacidcomposition, %							
	Lauric C 12:0	Myristic C14:0	Palmitic C16:0	Palmitoleic C16:1	Stearic C18:0	Oleic C 18:1	Linoleic C18:2	Linolenic C18:3
PI (dateofexamination 15.05.19)	0.2	0.9	38.5	0.3	6.3	37.9	15.5	0.4
PII (dateofexamination 13.04.19)	0.2	0.9	39.9		6.8	34.7	16.7	0.8

The fatty acid composition of the studied samples of potato puree concentrates is identical, probably this is due to the same raw material used in the technology by the same producer.

According to the fatty acid composition of the product, it follows that manufacturers choose palm oil rich in saturated fatty acids. First of all, this helps to increase the shelf life of the product, since palm oil is more oxidatively stable. At the same time, this does not have the best effect on consumer health.Excessive consumption of saturated fats is one of the most important risk factors for obesity, diabetes, cardiovascular and other diseases. Norms of consumption for saturated fats are not established by regulatory documents, however, according to WHO recommendations, their energy value in the norm of the total amount of fats should not exceed 10%.

Analyzing the data of Table 8.8, we can conclude that the peroxide value that is the safety indicator standardized by TR CU 021/2011 "On food safety" is within the normal range for all samples.

Table 8. Safety indicators of the fat fraction of instant potatoes (Personal research data)

Sample	Acid number, mgKOH/g	Peroxide number, meq½O/kg	Mass fraction of oxidation products insoluble in petroleum ether, %	Epoxycontent, meq/kg
PI (dateofexamination 15.05.19)	1.9	4.8	1.4	18.3
PII (dateofexamination 13.04.19)	1.6	3.9	1.3	21.1

The limit value of the acid number for fats is determined by the regulatory documentation of the Customs Union. According to TR CU 024/2011 "On fat-and-oil products". For refined oils and their fractions, mixtures of refined oils, the acid value is normalized to not more than 0.6 mg KOH / g. According to the results of the study, the acid number in all the studied samples is higher than normal and is 1.9 and 1.6 mg KOH / g, respectively, for PI- and PII-samples.

The mass fraction of compounds insoluble in petroleum ether is regulated in accordance with SP 2.3.6.1079-01. For used deep frying fats it is not more than 1%. This indicator has not yet been reflected in ensuring the safety of the fat component of other food products, although there are a number of works indicating the need for its regulation (Simakova I.V. et al.,2016a; SimakovaI.V. et al.,2016a).

In PI- and PII-samples, the mass fraction of compounds insoluble in petroleum ether is more than 1% - 1.4 and 1.3%, respectively. The content of epoxides is at the upper threshold of the recommended permissible level of 17-24 meq/kg and amounts to 8.3 and 21.1 meq/kg for PI- and PII-samples, respectively.

8.3 Instant Noodles

8.3.1 Technology for The Production of Instant Noodles

The general technology for the production of instant noodles consists of the following stages:

- doughing;
- forming of noodle products;
- heat steam treatment;
- drying;
- dehydrogenization in deep frying fat (blanched by fat) - dehumidification;
- packaging.

The detailed description of the process of production of instant noodles in different countries is presented below:

1. Doughing.

This is a process in which a blending machine mixes wheat flour and water.

2. Belt forming of the noodle.

The dough passes through two rotating rollers that mix 2 noodle belts into one. This process helps to spread the ingredients evenly. Sometimes the dough is left for maturation.

3. The noodles pass through four pairs of rollers and gradually become thinner. This process s gives the noodles elasticity.
4. Heat steam treatment. Parboiling.

During 1-5 minutes at 100 ° C.

5. Steamed noodles is dipped in spices.
6. Forming.

The noodles are cut at 40-70 cm in length and placed in a round or square shaped serving of metal mold.

7. Dehumidification.

The noodles in a roasting tin pass through the frying oil at 140-160 °C for a minute or two. The moisture content of 30-40% in the test is reduced to 3-6%.

Unfried noodles: the noodles in a roasting tin are placed in an air dryer and dehydrated with hot air at approximately 80 °C for more than 30 minutes.

Parboiling and frying in the deep-frying system ensures almost complete readiness and quick boiling of products.

8. Cooling.

After the dehydration process, air cooling follows.

9. Packaging.

Ready noodles are hermetically packed in film bags or in glass containers, together with spices and side dishes, which are then sealed with aluminum lined cap.

Wheat flour with a high protein content is the best raw material. It is carefully sieved, a certain amount of water is added and a thick dough is kneaded.

The composition, nutritional and energy value of noodles cooked according to the traditional recipe and instant noodles are presented in Tables 8.9 and 8.10.

Table 9 indicates that the richest composition of excipients by the number of ingredients is marked in instant noodles.

In addition to the above-mentioned substances that form taste and aroma, a number of other additives are used in instant noodles.

This may be such stabilizers as potassium carbonate E501 and sodium polyphosphate E452.

The main property of potassium carbonate is stabilizing (the ability to create and fix the necessary shape, texture and structure of the product), regulating the acidity of the medium (potassium carbonate is used to establish and maintain a certain pH level). It can also be used as an emulsifier and baking powder. In addition, the substance has an antimicrobial and keratolytic effect. Being an element vital for the normal functioning of the body, potassium is contained in the human body, and it can be washed out in the process of the human's vital activity. Potassium carbonate can be considered as a source of potassium replenishment. The substance is absorbed by the walls of the

Table 9. Analysis of the composition of instant noodles according to the label of this type of product (Personal analytical data)

Sample	Manufacturer / accessory	Composition	
		Mainingredient	Excipients
LB	Russia / International group of companies	Premium wheat flour, vegetable oil, water, tapioca and corn starch, salt. Sauce: drinking water, vegetable oil, sugar, salt, dried chicken meat, dried vegetables (onions, garlic, dill) Dried vegetables: corn, peas, carrots, scallions	The thickener - guar gum. Flavor intensifier - monosodium glutamate, spices (white and red pepper, turmeric), acidity regulators (acetic and lactic acids), stabilizer (guar and xanthan gums, egg yolk), preservatives (E202, E211), antioxidant (E385), flavoring agent (chicken), dye - beta-carotene, texturedsoyprotein
LD	Russia / Korea	Premium wheat flour, vegetable oil, salt. Seasoning broth: salt, broth base (chicken broth powder, soy sauce, garlic, dry spice concentrate, ginger, black pepper), sugar. Seasoning from dried vegetables: textured soy protein, carrots, onions, dried seaweed (cereal). The product contains gluten, soy and its processed products. May contain trace amounts of milk processing products	The thickener - starch, emulsifier "K-300A" (soybean oil, emulsifier lecithin), gluten, complex food component stabilizer "Premix" (potassium carbonate stabilizer, guar gum stabilizer, sodium polyphosphate stabilizer), mixed seasonings (sugar, yeast extract, vegetable protein, onion, soy sauce powder), dried kelp (powder), dried onion. Taste enhancer (monosodium glutamate), flavorant "Chiken", maltodextrin.
LR	Russia / International group of companies	Premium wheat flour, vegetable oil, salt, Seasoning broth - salt, sugar, vegetable oil, dried vegetables (garlic, paprika powder, onions, dill, tomato). Dried beef meat spices (red pepper, black pepper, thyme, nutmeg, bay leaf, clove). Dried vegetables (carrots, onions, textured soy protein)	Potato starch, egg powder, natural dye - beta-carotene. Maltodextrin, flavor intensifiers - glutamate, guanilate, sodium inosinate), flavorings identical to natural (beef)
LBL	Russia	Premium baking wheat flour, palm oil, starch, water, salt. Soup base: salt, sand sugar. Vegetables: scallions, onions, carrots. Chicken sauce: palm oil, fresh onion, chicken fat, salt, chicken fillet, fresh garlic, tomato paste	Flavor intensifier - sodium glutamate, onion powder, flavorant identical to the natural "Chicken", a mixture of spices - curry, ground black pepper, garlic powder, ground turmeric, acidity regulator - citric acid
LA	Russia	Premium wheat flour, durum wheat flour, palm oil, salt, sugar, carrots, boiled-dried chicken mince, parsley, onion, dill, garlic, red and black pepper. Dried vegetables: soya mince, red paprika, scallions, carrots, corn, green peas	Complex food supplement: salt, flavorant identical to natural, flavor enhancers (sodium glutamate, ribotide), turmeric. Spices: iodized salt

Continued on following page

Table 9. Continued

Sample	Manufacturer / accessory	Composition	
		Mainingredient	Excipients
LK	Russia / Korea	Premium wheat flour, vegetable oil, salt, broth seasoning: broth base (soy sauce, beef powder, garlic, radish powder, hot peppers, salt, sugar. Driedvegetablesseasoning: onion, garlic, mushrooms, redpaprika	Modified potato starch, gluten, emulsifier K-300 (soybean oil, lecithins), complex food supplement, stabilizer "Premix" (potassium carbonate, guar gum, sodium polyphosphate), mixed seasonings (sugar, yeast extract, vegetable protein, onions, soy sauce powder, flavorant "Beef", flavor enhancer (monosodium glutamate)

digestive tract; the remains are excreted along with the waste products.On the other hand, an excess of E501 can provoke the development of hyperkalemia, that is, a condition where the kidneys cannot remove all excess potassium from the body, and it begins to accumulate. It is the kidneys that suffer in the first place, since as a result, adrenal function may be impaired (Addison's disease).

Despite the fact that E452 polyphosphates are widely used in various types of production, are in daily contact with humans and the environment, they are only conditionally safe substances. Their use in food is permissible, but only in limited quantities. On average, the maximum amount of supplements consumed per day should not exceed 70 mg per 1 kilogram of body weight.

However, results of studies of polyphosphates are rather unsettling: they indicate that the use of polymers of phosphoric acid has a negative effect on the human body, as it contributes to the accumulation of phosphates in the body. Phosphates are carcinogens and provoke the development of tumors.

In addition, polyphosphates affect the assimilation of cholesterol in the blood, thereby causing vascular and cardiac diseases. The constant use of a stabilizer in a dosage exceeding causes problems with digestive function, dehydration of the body, and metabolic disorders in the cells.

There is evidence of a possible development of diarrhea, intestinal dysfunction, and vomiting in case of polyphosphates overdose. Given the relative safety of the substance, its amount in human food is strictly regulated by the legislation of the European Union, the Customs Union and the norms of each specific country at the domestic level.

Potassium sorbate (E202) and sodium benzoate (E211) are added to instant noodles as preservatives

Potassium sorbate (E202) is an organic compound. It is a potassium salt of sorbic acid. It is widely used in the food industry because of its ability to prevent the formation of mold and fungi. Its content is strictly regulated in food; it is not more than 0.02% of the total mass.The leader in the production of this preservative is the Chinese chemical concern Suzhou Import and Export Co., Ltd. The second place also belongs to the Chinese manufacturer - Shandong Reipu Chemicals Co., Ltd. There is no evidence of toxicity, mutagenicity or carcinogenicity of E202. In the case of excess of the daily intake or at individual intolerance to sorbic acid, a mild allergic reaction in the form of irritation of the mucous tissues and skin is noted. Many scientists oppose the use of this preservative because of its antibiotic properties. When it enters the human intestinal tract, the supplement destroys the beneficial microflora.

Sodium benzoate (E211) inhibits the vital activity of molds and yeasts, reduces the activity of microbial cells of enzymes that split starch and fats, and control redox reactions. Similarly, E211 affects the cells of the organs of the human body. At increased doses, sodium benzoate is a potent carcinogen. Today it has been reliably established that sodium benzoate provokes allergic reactions and causes the development of cancer.

The permissible daily intake of sodium benzoate is 5 mg/kg of body weight. In European countries, the harm of the combined use of the preservative of sodium benzoate and artificial dyes (E102, E104, E110, E122, E124, E129) is being studied. Based on obtained results, it has been made a decision to prohibit these dyes. The use of sodium benzoate is banned in some countries of the world (USA, Japan, Latin American countries).

Maltodextrin is a starch processing product. In food technology, it performs a number of functions: creates the effect of product thickening; allows one to save the color of the product as a result of inhibition of the oxidation process. It is a baking powder for dough and is used as an emulsifier. However, it has a high glycemic index, and for type 2 diabetics can be dangerous only during remission. Maltodextrin also suppresses the growth of probiotics.

Dry gluten is added to low-quality flour to produce flour that meets standard requirements. In European countries, the addition of gluten to weak flour is due to savings, because strong wheat is expensive and is usually imported from the USA and Canada. Gluten can be dangerous in the case of gluten enteropathy.

Due to the large number of excipients that have a number of technological advantages, but are not always safe for health, the consumption of instant noodles can be permissible only for healthy people.

Table 10 presents data on comparative analysis of the nutritional and energy value of instant noodles.

Table 10. Comparative analysis of the nutritional and energy value of instant noodles (Personal analytical data)

Sample	Manufacturer	Nutritionalandenergyvalue
		Per 100 g of dry product
Лапша яичная	According to the traditional recipe	protein -10.5 fat - 2.1 carbohydrate - 62.1 energy value - 310
LB	Russia / International group of companies	protein -8 fat -20.3 carbohydrate - 52.3 energy value - 424
LD	Russia / Korea	protein -8.7 fat -15.3 carbohydrate - 44.3 energy value - 349
LR	Russia / International group of companies	Белки - 7.7 fat -19.5 carbohydrate - 56.1 energy value - 431
LBL	Russia	protein - 9 fat -13 carbohydrate - 53 energy value - 360
LA	Russia	protein – 7.5 fat -20 carbohydrate - 49 energy value - 410
LK	Russia / Korea	protein - 9 fat -14 carbohydrate - 66 energy value - 420

The above analysis of the nutritional value of noodles requiring heat treatment and analogues that do not require boiling (instant noodles) shows a significant predominance of the fat component in all types of instant noodles. The amount of fat in instant noodles is 6.2 - 9.5 times greater than the amount of noodles requiring boiling; the energy value is also increased.

8.3.2 The Study of Safety Indicators of The Fat Component of Instant Noodles

In noodle samples, a comparative assessment of the mass fraction of fat extracted from the product and stated on the package was carried out (Figure 3).

Figure 3. mass fraction of fat in instant noodles, %(Personal research data)

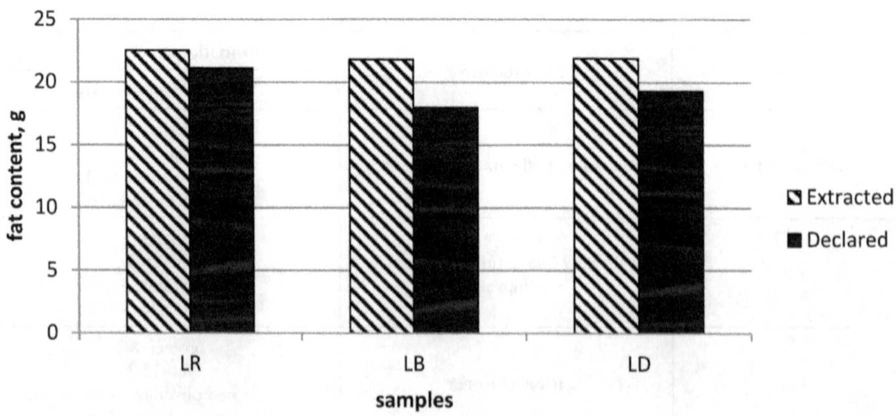

According to the analysis of the obtained results, the content of the mass fraction of extracted fat in all samples exceeds the declared on the package. A higher proportion of fat in foods is likely due to increased absorption of fat by the product during dehydration with hot oil.

According to the number of studies (Blumenthal, M.M. 1991; Bouchon, P., and D.L. Pyle. 2005a.), the increase in oil absorption depends on the state (quality) of fat. The influence of fat quality is explained by oxidative degradation of oils and the formation of compounds that increase the polarity of the frying medium. During the frying process, fat accumulates polar compounds, the level of surfactants increases, oil absorption proceeds more intensively. Although there are a number of other theories regarding oil absorption (Dana, D. 2006.), nevertheless, it can be noted that the rate of fat absorption can vary under the influence of numerous factors affecting the technological process of deep-frying. In other words, there is no possibility to predict the real nutritional value of the diet that includes these products. The fatty acid composition of the studied products is presented in Table 11.

Table 11. The fatty acid composition of fats in instant noodles (Personal research data)

Sample	Thefattyacidcomposition, %						
	C12:0	C14:0	C16:0	C18:0	C18:1	C18:2	C18:3
LR	0.4	1.2	44.6	4.3	38.6	10.7	0.4
LB	0.2	1.0	43.9	4.4	39.7	10.8	–
LD	0.2	1.1	44.7	4.1	38.9	10.6	0.4

In the study of the fatty acid composition of the analyzed products, it was found out that palm oil is prevailing in the fat component. Palm oil differs from other vegetable fats in its high content of saturated fatty acids. Excessive intake of saturated fats is one of the most important risk factors for obesity, diabetes, cardiovascular and other diseases.

Norms of consumption for saturated fats are not established by regulatory documents, however, according to WHO recommendations, their energy value in the diet in the norm of the total amount of fats should not exceed 10%. On the other hand, palm oil and its fractions can be considered as a partial alternative to hydrogenated fats in the production of certain types of oil and fat products, in particular, of deep frying fats.

Table 12 presents the quality and safety indicators of the fat component in instant noodle samples.

Table 12. The results of the study of safety indicators of the fat component in instant noodles (Personal research data)

Sample	Acid number, mgKOH/g	Peroxide number, meq½O/kg	Mass fraction of oxidation products insoluble in petroleum ether, %	Epoxy content, meq/kg
LR	0.3	6.0	1.2	52.1
LB	0.5	5.1	0.8	52.9
LD	0.3	5.5	1.1	52.3

Analyzing the data of Table 12, we can conclude that the safety indicator (peroxide number) is within the normal range for all samples, standardized according to TR CU 021/2011 "On food safety".

The limit value of the acid number for fats is determined by the regulatory documentation of the Customs Union. According to TR CU 024/2011 "On fat and oil products", the acid value or refined oils and their fractions, mixtures of refined oils, is normalized to not more than 0.6 mg KOH / g. According to the results of the study, the acid number does not exceed this indicator; only in sample 2 "the content of free fatty acids in the fat component" is close to the threshold value and is 0.5 mg KOH / g.

In accordance with SP 2.3.6.1079-01, the mass fraction of compounds insoluble in petroleum ether is regulated for used deep frying fats is not more than 1%. This indicator has not yet been reflected in ensuring the safety of the fat component of other food products, although there are a number of works that indicate the need for its regulation (Olijnik, Ja.V.,2007; Novozhilova G.P. et al.,2019).In samples 1 and 2, the mass fraction of compounds insoluble in petroleum ether is more than 1% - 1.2 and 1.1, respectively. The content of epoxides exceeds the recommended allowable level of 17-24 meq/ kg by almost 2.0-2.2 times.

Based on the results of the study, in the production of instant noodles, manufacturers comply with the safety parameters of deep frying fat regulated by TR CU 021/2011 and 024/2011. The need to comply with the requirements of SP 2.3.6.1079-01 is not taken into account.

In the normative documentation and in the real technological process, there is no control of epoxides, that are non-normalizable, but important oxidation products that affect human health. The formation of highly polar compounds insoluble in petroleum ether and epoxides was discovered by the authors in the study of deep-fried dough products, as well as during the storage of confectionery products with a high fat content.In a biological experiment on animals in our previous studies, it was shown that fats containing more than 1% of compounds insoluble in petroleum ether, have a negative effect on the organs of the gastrointestinal tract, sharply reduce the level of red and white blood cells, cause the accumulation of cholesterol and bilirubin (Simakova I.V., 2004).Our studies correlate with the studies of scientists from different countries. For example, there are data on pathological changes in the body when consuming oxidized and thermally oxidized fats (Fukuzumi, K., 1978). Fat decomposition products are toxic, they have a carcinogenic effect, can cause diseases of the gastrointestinal tract and liver, growth retardation, irritation of the skin and mucous membranes, and the development of malignant tumors (Vysheslavova, M.Ja., 1969; Sanders, T.,1987; Carroll K.K., 1989).

The study of the safety of the fat component of instant concentrates - soups, potato puree, and noodles revealed found that the safety parameters of the fat component regulated by TR CU 021/2011, and TR CU 024/2011 do not fully reflect the safety requirements for the fat component of fast food blanched with hot oil, since there are no standards for the most important indicators of fat safety - the content of secondary oxidation products and epoxides. In our opinion, studies lead to the need for further study of safety indicators for the fat component of fast food products in terms of their effect on the body when they are included in diets.

Obtained data indicate the need for a critical assessment of production technology, as well as the advisability of introducing regulatory standards regarding the safety control of the fat component of fast food products, in order to reduce the effects of toxic products of fat oxidation on the human body and, as a consequence, reduce the development of the burden of civilization diseases.

8.4 Features of Technology and The Effect of Kissel On the Human Body

In everyday human nutrition, including the nutrition of children, especially in social institutions, dishes and products according to traditional technology are replaced by concentrates of industrial production. At the same time, many authors (Kodensova V.M. et al., 2017; Migunova Yu. V., Sadykov R. M., 2018), including studies by the Federal Research Center for Nutrition and Biotechnology of the Russian Federation (Methodical recommendations MR 2.3.1.2432-08 "Norms of physiological requirements for energy and nutrients for various groups of the population of the Russian Federation": approved by Chief State Sanitary Doctor of the Russian Federation 12/18/2008.), as well as data from the World Health Organization, note a change in nutrition status, in the direction of essential substance deficiency. Therefore, the determination of the effect of individual food products, especially from concentrates on the body is relevant.

Concentrated kissels and extended-shelf prepared foods became more and more popular. Concentrated kissels are similar in nutritional and energetic value to dishes according to traditional technology. However, it should be noted that a significant difference in the balance of carbohydrates by their type can be observed in them. In addition, these products contain extracts of fruits and berries, a number of food additives (dyes, flavors, etc.), which cannot be equivalent to a natural raw material component.

In the scientific literature, the qualitative indicators of gelled dishes are characterized mainly by their structural and mechanical properties: the size of microgels, the degree of swelling and temperature sensitivity in aqueous solutions, as well as the possibility of using glucose syrup concentrates in the nutrition of newborns ill with hypogycemia, etc. (Gau E. et al,2017; Makker K. et al., 2018).

However, there are no studies on the establishment of safety criteria for concentrates, including concentrates of desserts, on the appropriateness assessment from a physiological point of view, and their use in social nutrition.

The purpose of the work is to study the quality indicators of concentrates of instant desserts.

The objectives of the work:

1. to determine the functional and technological properties of starch polysaccharides of kissel concentrates from natural raw materials that do not require heat treatment;
2. to conduct a comparative assessment of kissel concentrates from natural raw materials with analogue ready meals prepared according to the Collection of technical standards;
3. to study the effect of kissels from industrial production concentrates in comparison with kissels prepared according to the Collection of technical standards on the morpho-functional parameters of the organism of experimental animals.

Research methods. The subject of research was the organoleptic physico-chemical and biological indicators of the dessert concentrates. The objects of the study were fast food products, the most popular among consumers in the Russian Federation - kissels of popular brands:

- kissels of the "Compass Zdoovya" brand, consisting of natural raw materials (oat flour, oatmeal and other natural ingredients);

  - the product "Oatmeal breakfast" of the Velle trademark; its composition and organoleptic characteristics are the closest to a thick kissel, an analogue of authentic Russian cuisine;
  - ready-made kissels of the "PROSTO" brand;
  - cranberry kissel (No. 1082 according to the Collection of Technical Standards. Collection of recipes at public catering enterprises, 2016) and similar kissels from concentrates of Russian and Polish companies in a labile price category.

To study the properties of starch polysaccharides, microscopy was used using a fully functional stereo microscope with an additional channel for recording images through a video eyepiece.

Vitamin C was determined by the titrometric method according to GOST 24556-89.

Organoleptic evaluation of kissels was carried out by the profile method in accordance with GOST R ISO 5492-2005 and GOST ISO 13299-2015. Organoleptic analysis.Methodology.General guidance for the compilation of an organoleptic profile.

Statistical processing of the described studies complies with international requirements and is carried out in accordance with ISO 5725 (parts 1-6) and ISO / IEC 17025 - 99. All test results are processed by standardized measurement methods and meet the requirements of trueness and precision of measurements. The replication of measurements is at least 5 times, a confidence probability of the accuracy of test results is 95%.

8.4.1 Technological Features of The Production of Concentrates of Desserts – Kissels

Historical note. Grain-based kissels existed 9,000 years ago in ancient Anatolia and Mesopotamia, and are mentioned in Sumerian and Akkadian texts. Oatmeal-based kissel soup (zurek soup) is one of the characteristic national dishes of Polish and Belarusian cuisines.

In ancient times, to produce jelly oat milk was fermented (by mixing oatmeal with water or by pressing soaked cereals).

Oatmeal, rye, wheat kissels belong to the most ancient Russian dishes. Kissels were an everyday dish, as well as a ritual dish for funeral feasts. In hot form, oatmeal kissel was more often eaten with linseed or hemp oil, and cooled and thickened kissel was cut with a knife and eaten with milk, jam or fried onions.

Sweet fruit and berry kissels, which appeared later in the 19th century, when potato and cheap potato starch spread in Russia, inherited this name. Their wider introduction was also facilitated by their simpler formulation and short cooking time, due to the lack of fermentation during cooking. In European countries, the word "kissel" is borrowed from the Russian language. For example, in Estonian language – kissel, in Latvian – ķīselis, in Lithuanian - kisielius, in German language - kissel mean oatmeal kissel. Fruit and berry sweet kissel based on starch do not have a common name in any Western European language (attributed to fruit cereals, gravy, sautés, fruit sauces, etc.).

Gelatinous dishes, like kissels, are widespread in Western European and world culinary. A striking example is the rice pudding, found around the world in a variety of variations. However, the proximity of recipes is equally characteristic for oat, pea, milk and berry-fruit kissels, which is natural with close trade and cultural exchange.

A fairly accurate analogue of grains flour kissels can be found in British cuisine of the 17th – 19th centuries - flummery. This dessert was cooked from soaked oat or wheat seeds, but without fermentation. It was served with honey, cream and other additives. The presence of a fermentation stage in the

Russian tradition is noteworthy, since our cuisine as a whole is characterized by an acidic taste.

Flummery is considered a variety of puddings, popular in English cuisine. Also in Great Britain there was an analogue of the Russian salamata - gruel. It was the main dish of the workhouse inhabitants in the novel “Oliver Twist” by Charles Dickens.

The German counterpart to oatmeal kissel is Haferschleim - oat broth, porridge. In addition, in German and Danish cuisine there is a dish completely similar to kissel from potato starch (German – roteGrütze, Danish rødgrød). This sweet dessert with red summer berries was originally cooked from cereals, then potato starch was used as a thickener. Rote Grütze is also served chilled with milk or cream.

In French cuisine, berry-fruit kissels, which were prepared with the addition of gelatin based on fish waste, are closest to kissels from starch. In the “Gastronomy Almanac” (1852–1855) by Ignatius Radetzky, where Russian-French cuisine of the mid-19th century is presented, the names of kissels are duplicated in French as “gelèe (kissel)”. However, Radetzky does not mix these dishes: the book contains recipes for raspberry and cranberry kissels and jelly from the same berries.

The Turkish delicacy Lokum, which is cooked from starch with pink water, mastic tree resin or fruit juices as the main flavoring essences, is similar to kissels from potato starch. An analogue of pea kisselis easily found in Italian cuisine - it is a polenta cooked from cornmeal (mamalyga in Eastern-Roman countries).

Thus, it can be noted that kissels or their analogues are characteristic of many countries of the world.

There are several types of dessert concentrates:

- kissels, mousse, jelly, creams, puddings.
- dessert food concentrates.

Dessert food concentrates include a group of concentrates, which are a mechanical mixture of granulated sugar, potato or corn starch, semolina, wheat flour, milk powder and various flavors included in a product in the amounts established by the recipe.

According to the method for producing a ready dessert, food concentrates can be divided into two groups:

- concentrates requiring boiling, which are a mixture of granulated sugar, potato or corn starch, semolina, wheat flour, milk powder and various flavors;
- concentrates that do not require boiling (instant cooking), consisting of a mixture of fruit puree, potato starch and sugar, dried and packaged in consumer packaging made of plastic film.

Concentrates requiring boiling include:

1. concentrates on fruit or berry extracts or concentrated juices:
    a. kissels (cherry plum, lingonberry, cherry, blueberry, blackberry, cornel, strawberry, cranberry, raspberry, rowan, plum, blackcurrant, thorny, blueberry, apple, fruit and berry, chokeberry and rowan);
    b. mousse (lingonberry, cherry, strawberry, cranberry, raspberry, plum, blueberry, blackcurrant, apple, fruit and berry);
    c. jelly (orange, cherry, strawberry, cranberry, lemon, raspberry, tangerine, blackcurrant, fruit and berry, "Novinka" cherry;
2. milk concentrates:
    d. kissels (milk, milk chocolate);
    e. jelly creams (vanilla, coffee, milk, chocolate);
    f. custard creams (custard, coffee, cream, chocolate);
    g. dessert puddings (orange, vanilla, coffee, lemon, almond, chocolate).
        2. Concentrates that do not require cooking include jelly on applesauce.

All types, depending on the raw materials used, are divided into:

- concentrates of sweet dishes on fruit or berry extracts of concentrated juices, (jelly, mousse, jelly);
- Concentrates that do not require boiling include kissels from applesauce.
- Depending on the raw materials used all kissels are divided into following types:
- dessert concentrates from fruit or berry extracts of concentrated juices (kissels, mousse, jelly);
- dessert concentrates without fruit and berry extracts of concentrated juices (with flavoring and aromatic additives - kissels, mousse, jelly);
- dairy dessert concentrates (kissels and jelly and custard creams);
- dessert concentrates without dairy components (creams and dessert puddings).

Dessert food concentrates requiring boiling. Concentrates are produced according to the following technological scheme: preparation of raw materials, dosing and mixing, packaging.

Raw material preparation has the most significant effect on the quality of food concentrates.

In the production of kissel food concentrates, starch or dry fruit semifinished product enters the borate through a centrifugal unloader. Control screening of the product is carried out on the borate. Before sifting, potato and gelling starch is dried to moisture content of 10 ... 12% to avoid lumpiness.

The temperature during the drying of starch should not exceed 40 °C, since a higher temperature can lead to a change in the physicochemical properties of starch and, consequently, to a deterioration in the swelling and consistency of ready dishes. For starch sifting, metal-woven sieves No. 1.2 ... 1.6 are installed, and for dry fruit semi-finished products sifting sieves No. 2 ... 2,5 are installed. The product enters a standardized dispenser from the sifter, and flows to a continuous mixer.

Granulated sugar is sifted on a sifter on which a metal-woven sieve No. 2 ... 2.5 is installed, and then it is fed through a centrifugal unloader to a unified dispenser, and from a dispenser to a mixer.

When preparing kissel from fruit and berry extracts, the corresponding extract also enters the continuous mixer. From the reserve tank it is fed to a blending tank equipped with a mixer. If necessary, citric acid which dissolves in the extract is added to the blending vessel. In this case, it is necessary to monitor the complete dissolution of the acid in the extract. If the acid is not completely dissolved, defects may form on the surface of the kissel briquettes. These defects are white spots due to the discoloration of the extract by crystals of citric acid that dissolve in this place.

The extract with citric acid dissolved in it is fed through a filter by a metering pump to a continuous mixer. The extract is filtered through a metal-cloth sieve No. 1 or silk sieve No. 15. From the mixer, the mass of kissels goes to the briquette press.Briquettes formed on the press are directed to wrapping machines, where they are packed in parchment or other material with a company label. Briquettes are placed on a table in a shipping container - corrugated cardboard box. Boxes are glued over with a parcel on a banding machine, then they are directed to the finished product warehouse.

Fruit and berry kissels can also be produced non-briquetted, in packages of kraft paper weighing up to 3 kg for a catering network. For individual consumption, they are packed in polymeric bags and have smaller weight.

In the production of kissels from dry fruit semi-products, only sugar and dry fruit semi-product are set in the mixer. A well-mixed mixture is directed to a filling machine, packaged in polymeric bags or in paper coated with polyethylene, weighing 33 ... 250 g, without briquetting.

Table 13. The composition of instant kissels (Personal analytical data)

Product	Manufacturer	Composition	
		Main ingredient	Excipients
Kisselfromfruits	Traditional recipe	Potato starch, cranberry, blueberry, sugar, water	-
Fruit kissel	PAO "Russky Product"	Potato starch, concentrated apple and chokeberry juice	Lemon acid; sugar; flavorant; vitamin premix; dye - concentrated carrot juice
Instant kissel	OOO"Pets-haas"	Potato starch, sugar	Lemon acid; salt; natural flavorant "Strawberry"; dyes: carmine, annatto
Kisselwithraspberryaroma	Agrippina	Corn starch, potato starch; sugar; natural apple extract or concentrated apple juice	Food lemon acid monohydrate; carmine dye; flavorant "Raspberry", vitamin premix
Cranberrieskissel	PAO "Russky Product"	Potato starch; sugar; natural concentrated cranberry juice	Lemon acid; flavorant "Cranberry"; dyes: E102; E122; E133
Cherryoatmealkissel	Fitparad	Oatflour; cherrypowder	Maltodextrin; pectin; inulin; lemon acid; Steviosin sweetener; carmin
Detox kissel "Lose Weight in a Week"	OOO "Leovit"	Starch; sugar; oatmeal; fruits and berries; beet	MgSO4; lemon acid; guar gum; celery; corn stigma extract; coriander; flavorant; bay leaf; green tea. Product may contain traces of peanuts, nuts, sesame seeds, celery, cereals with gluten, milk and / or their processed products
Vitamin kissel for eyes with lutein	OOO "Leovit"	Potato starch; sugar; fruits and berries; vegetables	Lemon acid; guar gum; oat flour; vitamins (C, E, rutin); carotenoid lutein; flavorant; royal jelly (extract)
Instant kissel	PAO "Russky Product"	Potato starch; sugar; natural dried berries	Lemon acid; flavorant; vitamin premium; natural dye; anthocyanins
Fruit kissel	"Babushkin hutorok"	Corn starch; sugar; potato starch; natural concentrated juice	Lemon acid; flavorants "Apple", "Cranberry"; vitamin C
Blueberrykissel	"Kulinar"	Starch; concentrated juice; sugar	Lemon acid; flavorant "Blueberry"

Tables 13 and 14 present the composition, nutritional and energy value for some kissels produced by different manufacturers and various cooking methods.

The composition of kissels from concentrates does not differ in variety (Table 13) in the main ingredients; starch, sugar, dried fruit or berries in powder, dried fruit or berry extract prevail, however, dried components of fruit and berry raw materials are often imitated by adding dye and flavorant.

According to Table 13, starch is the main kissel thickener, regardless of the basic raw materials.

Starch is not only a reserve polysaccharide in kissel, it has an enveloping effect on the walls of the gastrointestinal tract, contributing to the normalization of its work. Due to gelatinization during thermal exposure, amylopectin (a branched form of starch) forms a viscous, gelatinous consistency.This consistency normalizes the internal environment of the intestine, and starch delays the synthesis of secondary bile acids toxic to the body. Starch is considered to be an easily digestible polysaccharide, and the same amounts of fully adsorbed and stable starch can have different effects on glycemic and hormonal indices in people, causing a feeling of satiety.

The prototype of traditional kissel in authentic Russian cuisine was produced on the basis of oat flour, in the native composition of which was not only starch, but also mucus - complex mixtures of heteropolysaccharides. Therefore, the formation of functional and technological properties was influenced not only by technological methods, but also by constituent components.At present, the production of kissels differs from the original technology; pre-cooked, usually modified industrial starch (purified from mucus and associated interacting natural components) is used.

Digestion of starch starts in the oral cavity under the action of the saliva alpha-amylase enzyme. In the stomach, this enzyme no longer works, there is an acidic environment. The main place for starch digestion and glucose absorption is the small intestine, where the alpha-amylase enzyme and bicarbonate (the reaction of the medium is alkaline) come from the pancreas.

An indicator of the digestion of starch and absorption of the product of its enzymatic glucose hydrolysis is a sugar increase in blood after eating. There is evidence that the use of resistant starch can reduce the level of insulin and lipids in blood plasma, as well as the activity of lipogenic enzymes in the liver.

Fermentation of resistant starch in the colon is also very important for alternative excretion of urea nitrogen (up to 50%) from the body with feces. But, as it turned out, not every starch is digested.

Resistant starch is a starch undigestible by enzyme of the gastrointestinal tract. It reaches the large intestine, becomes food for fecal microflora and is fermented. Resistant starch fermentation products are short-chain fatty acids with various physiological and probiotic effects. The scientific interest in resistant starch has increased significantly in recent decades due to the possibility of the formation of high levels of butyrate (butyric acid) from resistant starch in the large intestine. Butyrate is the most important energy source for colonocytes - cells of the large intestine mucosa and has a beneficial effect on metabolism and cell growth. It reduces cholesterol, triglycerides and urea in blood, and also prevents a number of factors that contribute to the initiation, progression and growth of a colon tumor.

Type I (RS1) - physically nonavailable starch. Amylolytic enzymes do not have access to starch in native plant cells (grain), since there are no enzymes in the gastrointestinal tract that can decompose the components of plant cell walls. Such starch, together with fragments of plant tissue, enters the intestine.

Type II (RS II) - raw starch of certain plant species, for example, potatoes, green bananas.

Type III (RS III) - retrograde starch, i.e. spontaneously or artificially precipitated from starch paste in the form of semi-crystalline structures insoluble in water. As a result of retrogradation, amylose rather than amylopectin forms more thermostable structures. The amount of resistant starch increases along with an increase in the amylose content of starch.

Type IV (RS IV) - chemically or physically modified starch.

It is believed that the optimal daily dose of resistant starch is 15-30 g.

Foods high in resistant starch

- Oats

100 g of boiled oatmeal may contain 3.6 g of resistant starch. In cooled oatmeal, the content of resistant starch increases. This also applies to the technology for the production of traditional oat kissel.

- Cooked and chilled rice

Brown rice is preferred over white rice. Brown rice also contains more trace elements, including manganese and magnesium.

- Legumes

Beans and legumes (lentils, chickpeas, peas, beans) provide a large amount of dietary fiber and resistant starch. To remove lectins and inhibitors of digestive enzymes they must be kept in water for several hours and well boiled. Depending on the type of legumes, in prepared form they contain approximately 1-4 g of resistant starch per 100 g,

- Potato starch

Potato starch is a white powder similar to regular flour. This is one of the most concentrated sources of resistant starches, about 72% of potato starch is resistant starch.

- Cooked and chilled potatoes

Cooked and chilled potatoes are a good source of resistant starch. Chilled potatoes intended for cold dishes (as an ingredient in salads) contain significantly more resistant starch than hot potato dishes.

- Green bananas

Green bananas are another great source of resistant starch and fiber.

- Corn starch

Corn starch is a very concentrated form of resistant starch. It can easily be added to yogurt or oatmeal. Corn starch contains up to 50% of fiber, most of which is resistant starch.

High amylose corn starches are resistant to digestion in the small intestine. Therefore, resistant starches are also useful in the diet as low-calorie foods.

Based on the foregoing, it can be noted that desserts, such as kissels, as well as concentrates, can be a source of resistant starch.

In the scientific periodicals, relatively many works are devoted to the study of the effect of various types of resistant starch on the body (Sajilata M.G. et al., 2006; Zhou J. et al., 2008; Shen L., et al., 2011; Bodinham CL, et al., 2011; American Diabetes Association Classification and diagnosis of diabetes. Diabetes Care. 2017; Cho N.H. et al., 2018); it is indicated that resistant starch improved insulin resistance in patients with type 2 diabetes mellitus with obesity.A number of studies also contain information on the effect of resistant starch when consumed by people with cardiometabolic risk factors on the reduction of inflammatory markers TNF-α and heart rate, as well as on biomarkers for glucose and appetite regulation.

Based on the foregoing, it can be concluded that the consumption of kissels in the daily diet of a modern person is very useful.

Table 14 presents an analysis of the nutritional and energy value of instant kissels.

Table 14. Nutritional and energy value of instant kissels(Personal analytical data)

Product	Manufacturer, brand, production	Nutritional and energy value per 200 ml	Note
Kisselfromfruits	Traditional recipe	Carbohydrate - 19.2; energyvalue - 76.8; B1 - 0.006; B2 - 0.008; B3 - 0.06; B6 - 0.016; C - 3.16.	-
Fruit kissel	PAO “Russky Product”, Russia	Carbohydrate - 27.3; energy value - 109.8; B3 - 2.376; B6 - 0.289; C - 9.36.	90 g of concentrate per 600 ml
Instant kissel	OOO“Pets-haas”, Russia	Carbohydrate - 27.9; energyvalue - 112;	30 g per 250ml
Kisselwithraspberryaroma	Agrippina, Russia	Carbohydrate - 26.7; energyvalue - 102; B1 - 0.03; B2 - 0.03; B6 - 0.035; C - 1.6.	220 g per 1400 ml
Cranberrieskissel	PAO “Russky Product”, Russia	Carbohydrate - 27.3; energyvalue - 108.	220 g per 1300 ml
Cherryoatmealkissel	Fitparad, Russia	Carbohydrate - 12.88; protein - 0.93; fat - 0.44; energy value - 59.27; B1 - 0.03; B6 - 0.016; C - 0.83.	50 g per 600 ml
Detox kissel “Lose Weight in a Week”	OOO “Leovit”, Russia	Carbohydrate - 17.8; energy value - 68; Mg - 40.	20 g per 200ml
Vitamin kissel for eyes with lutein	OOO “Leovit”, Russia	Carbohydrate - 16; energy value - 65; E - 11.3; C - 45; rutin - 15; lutein - 2.5.	18 g per 200ml
Instant kissel	PAO “Russky Product”, Russia	Carbohydrate - 11.8; energyvalue - 47.5; B3 - 1.3; B5 - 0.72; B6 - 0.16; B9 - 0.03; C - 4.98.	150 g per 1200 ml
Fruit kissel	“Babushkin hutorok”, Russia	Carbohydrate - 18.4; energy value - 73.6; C - 10.	180 g per 1000 ml
Blueberrykissel	“Kulinar”, Russia	Carbohydrate - 13.2; energyvalue - 52.	220g per 1400 ml

According to an analysis of the nutritional value, kissels from concentrates differs in the amount of carbohydrates from the kissels according to the traditional recipe. Content of carbohydrates decreases from 19.2 g per serving to a minimum of 11.8 g per serving for instant kissel of PAO “Russky Product”; at the same time, content of carbohydrates increases in kissel, requiring boiling: fruit kissel of PAO “Russky Product” - 27.3 g, Kissel with raspberry aroma of Agrippina - 26.7 g per serving and instant kissel of OOO "Pets-haas" - 27.9 g per serving, respectively. The energy value of the studied concentrates varies from 47.5 to 109.8 kcal, when compared with the calorie content of Kelly according to the traditional recipe, when it is 76.8 kcal.

Fruit kissels are a source of digestible and non-digestible (resistant) starch and vitamins, while concentrates, unlike kissels according to the traditional recipe, have to be specially enriched with a vitamin-mineral complex. Table 15 presents the content of vitamin C in kissels. Cranberry kissels according to the traditional recipe and analogue from concentrates of different manufacturers were selected for the study.

Table 15. The content of vitamin C in kissels (Personal research data)

Sample	The content of vitamin C, mg per 100 g	
	theoretical	experimental, for the finished product
1 (control)	18	18.7
KI (Poland)	Notapplicable	0.6
K II (Russia)	Notapplicable	0.6
KIII (Russia)	100 (for 100 g of dry product)	1.8

It is obvious that the examined samples do not reach the established level of traditional kissel (control). Thus, kissels from concentrates may not always be a source of vitamins.

Further research was carried out in two directions:

- determination of the functional and technological properties of concentrates that mimic the initial formulation of oatmeal kissels and the identification of their possible impact on the human body;
- determination of the real effect of kissel concentrates on the body of experimental animals with regular consumption.

8.4.2 Functional and Technological Properties of Concentrates

In the study of the functional and technological properties of concentrates that mimic the initial recipe for oatmeal kissels, the preparation technology proposed by NPO "Compass Zdorovya" (Russia) is considered. A preliminary descriptive sensory assessment was performed followed by analysis of the microstructure of starch polysaccharides (Table 16).

Table 16. Organoleptic analysis of kissel samples of NPO "Compass Zdorovya", Russia (Personal research data)

Sample	Oatmealkissels	Carrot kissel»	Gingerkissel
Exterior	Beige-gray powder with a slight grayish blotches	Beige-gray powder with orange and yellow blotches	Beige-gray powder with with cream splashes
Aroma	Soft, slightly expressed, sweetish, moist, characteristic of the product	Pronounced, the flavor of oatmeal, a smell of carrots inherent in the product	Soft, with a pleasant smell of ginger, the smell of flour is felt to a lesser extent
Consistence	Inhomogeneous, with particles of different sizes	Inhomogeneous, with particles of different sizes, including particles of carrot	Inhomogeneous, with particles of different sizes
Consistenceof brewing kissel	Inhomogeneous, particles aggregate with each other, requires thorough mixing. Deposits, forming two fractions, viscosity does not meet established requirements	Inhomogeneous, particles aggregate with each other, requires thorough mixing. Deposits, forming two fractions, viscosity does not meet established requirements	Inhomogeneous, particles aggregate with each other, requires thorough mixing. Deposits faster, due to the presence of different-sized particles, forming two fractions, the viscosity does not meet established requirements; it has large lumps that are difficult to stir
Tasteandflavor	Taste of oatmeal, when consumed, a heterogeneous consistency is strongly tasted	Taste of oatmeal and carrot, when consumed, a heterogeneous consistency is strongly tasted	It has a pronounced pleasant taste of ginger, flour is tasted less than in previous samples

Based on the Table, it can be noted that the method proposed by the manufacturer does not allow forming a structure inherent in kissels. The required viscosity is missing. The dissolved product is more consistent with the suspension than with the gel characteristic of the kissel structure.This suggests that starch polysaccharides have not reached the gelatinization stage, and are probably at the swelling stage prior to gelatinization. Preliminary sensory assessment data suggested the impossibility of manifesting the therapeutic properties of the kissel in these concentrates.

To assess the possibility of structure formation, it was proposed to conduct heat treatment of concentrate kissels (heating to a temperature of 100 °C).

When heated to 100 °C and boiled for 1 minute, the onset of gelation characteristic ofkissel was observed, however, large particles did not dissolve, but only swelled. The homogeneous, mucous consistency inherent in kissel was not fully achieved.

The study of the properties of produced kissels, including microscopy of samples, showed that the preparation method proposed by the manufacturer does not form a viscous-plastic consistency characteristic of kissels. Most starch grains are in their native form and are in the swelling stage (Figure 4).

Partial color staining of the field around the grains in dark-blue and red-violet indicates the gelatinization of amylose and amylopectin in starch composition, and it should be noted that there are empty spots in the microscopic field. Figure 8.5 shows a kissel sample of NPO "Compass Zdorovya" after changing the cooking conditions (bringing to a boil after brewing). Obviously, the starch grains defined in the first sample are absent. Blue-violet staining indicates the formation of amylose-amylopectin complex, forming a viscous jelly. The presence of space in the field of view allows one to state that the system does not achieve uniformity and remains in an intermediate state between the suspension and the gel.

Figure 4. Microscopy of kissel of Compass Zdorovya brand prepared according to the method proposed by the manufacturer (Personal research data)

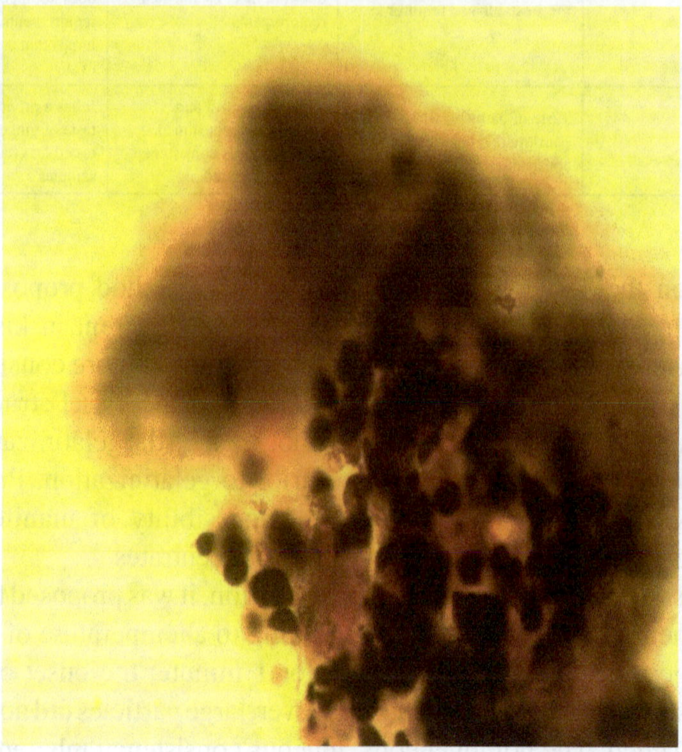

To compare the characteristics of starch polysaccharides, we performed microscopy of samples of finished products sold to the population through a retail network. As far as functional and technological properties are

Figure 5. Microscopy of kissel of Compass Zdorovya brand after bringing to a boil (Personal research data)

concerned, the oat breakfast of "Velle" brand (Russia) corresponds to thick kissels (Figure 6). The gelatinous consistency is visible in the structure of the product, with the predominant blue color characteristic of amylose. It should be noted that whole grains of starch of irregular shape are also traced. This indicates incomplete gelatinization, which may also affect the digestion of the polysaccharide as a whole.

The results of microscopy of the ready kissel of the "PROSTO" trademark are of great interest (Figure 7). The absence of gelatinization in general should be noted; starch grains are swollen, irregular in shape, while the release of amylose and amylopectin from the grain is not observed. Apparently, the initial stage of gelatinization can explain the low viscosity of the product. Based on the microscopic picture of the finished product, there is doubt about the possibility of the preventive effect of this product on the human gastrointestinal tract.

Figure 6. Microscopy of the oat breakfast of "Velle" brand (Personal research data)

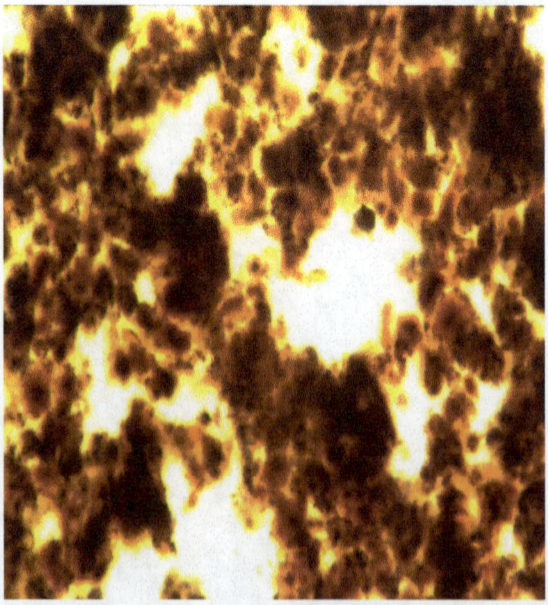

Figure 7. Microscopy of kissel of "PROSTO" brand (Personal research data)

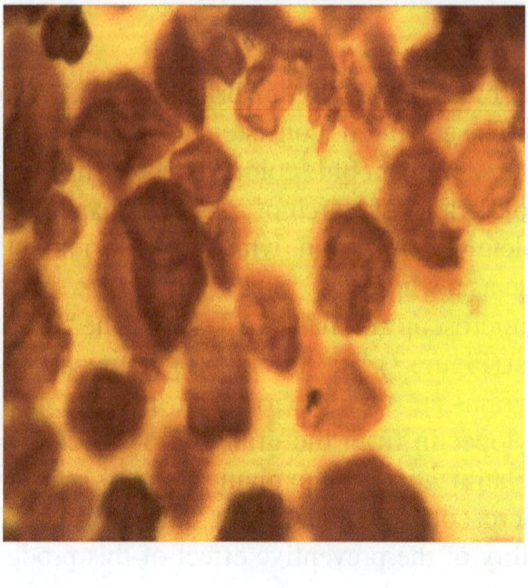

8.4.3. The Effect of Kissel Concentrates on The Body of Experimental Animals with Regular Consumption

All animal experiments were carried out on the basis of a certified vivarium of the educational and scientific-technological center "Veterinary Hospital" and laboratories of the department "Morphology, animal pathology and biology" of the Saratov State Agrarian University named after N.I. Vavilov. Studies were carried out in accordance with the "Rules for the work on experimental animals" (Guide for the Care and Use of Laboratory Animals (National Research Council, National Academy Press, 2011).

The animals were fed for 60 days; during the entire experiment, the rats were kept in individual cages (10 animals in each group). Prior to the introduction of the studied products into the diet, animals were kept in quarantine for 21 days and transferred to the diet in accordance with the research plan.The effect of kissels from concentrates on the body of animals in comparison with kissel according to traditional recipe, with prolonged consumption was studied by pathomorphological and histological research methods.

All experimental studies were performed on groups of clinically healthy rats of the same breed, one sex, one age, one weight. The animals were fed for 45 days; during the entire experiment, the rats were kept in individual cages (10 animals in each group).

The work was performed on four groups of clinically healthy rats (10 animals in each group), formed according to the principle of analogues taking into account breed, sex, age (1.5 months), live weight and clinical condition: the control group was fed with the usual full-fledged diet;the first experimental group was fed with a complete diet and kissel according to the traditional recipe; the second experimental group was fed with the usual high-grade diet and briquette concentrate kissel (Russia);the third experimental group was fed with the usual high-grade diet and powdered concentrate kissel (Poland). Replacement was carried out without prejudice to the nutritional and energy value of the daily diet of animals. Prior to the introduction of products into the diet, animals were kept in quarantine for 21 days and transferred to the diet in accordance with the experimental plan.

An autopsy was carried out with detailed recording and photographing of the material. Pathomorphological changes were studied on material from 40 euthanized animals (Recommendations for euthanasia of experimental animals: Part 1, Part 2. LaboratoryAnimals, 1996).

The autopsy of animals was carried out in the first 2 hours after the forced slaughter. For histological examination, pieces of the liver and spleen were taken. A 10% solution of neutral formalin was used to fix the pathological material.

Sections were produced on a freezing microtome of 2515 model (Reichert Wien). The prepared histological sections were stained with Ehrlich hematoxylin and eosin, followed by microscopy. The morphological structure of organs was studied in 30 fields of view of the microscope in various histological sections.

A histological examination of the preparations was carried out under different magnifications, with detailed recording and photographing of the studied areas. Microphotography of histological preparations was carried out using a CANON Power Shot A460 IS camera. Statistical analysis of these in vivo experiments was carried out using the licensed computer software package Statistika 10, StatSoft, Inc (Series 0411-R) using biomedical statistics methods with calculation of the mean and standard error of the mean.The normality of the distribution of quantitative data was checked using the Shapiro-Wilks test. For all indicators, including behavioral studies with a normal distribution of data, intergroup comparisons were performed using the Student t-test for independent samples at P <0.05 (S. Glants, 1998.)
Table 16 presents the dynamics of live weight of experimental animals.

Table 16. The dynamics of live weight of experimental animals (Personal research data)

Group of animals	The average weight before feeding, g	The average weight after feeding, g	Liveweight gain, g and %	
			g	%
Control	163.7±1.5	209.9±0.8	46.2	28.2
1 experimental	162.9±5.3	215.5±0.9	52.6	32.2
2 experimental	158.8±8.06	223.03±1.3	64.2	40.4
3 experimental	155.6±2.7	294.3±0.9	138.7	89.1

Table16 evidences that the animals of the first experimental group were slightly different in liveweight gain (4%) from the control. At the same time, in the second group, in comparison with the control and the first experimental group, there was an increase in liveweight by 12.2 and 8.2%, respectively. This is most likely due to the fact that simple sugars and starch are the prevailing carbohydrates in the briquetted concentrate (Russia).

Significant liveweight changes were noted in the third experimental group, and amounted to 89.1%, which is 60.9 and 56.9% more than in the control and in the first group, respectively. The results obtained allow one to make an assumption about impaired carbohydrate metabolism in rats of the third experimental group (Figures 8-11, Table 17).

Figure 8. The intestine of rats in the control group

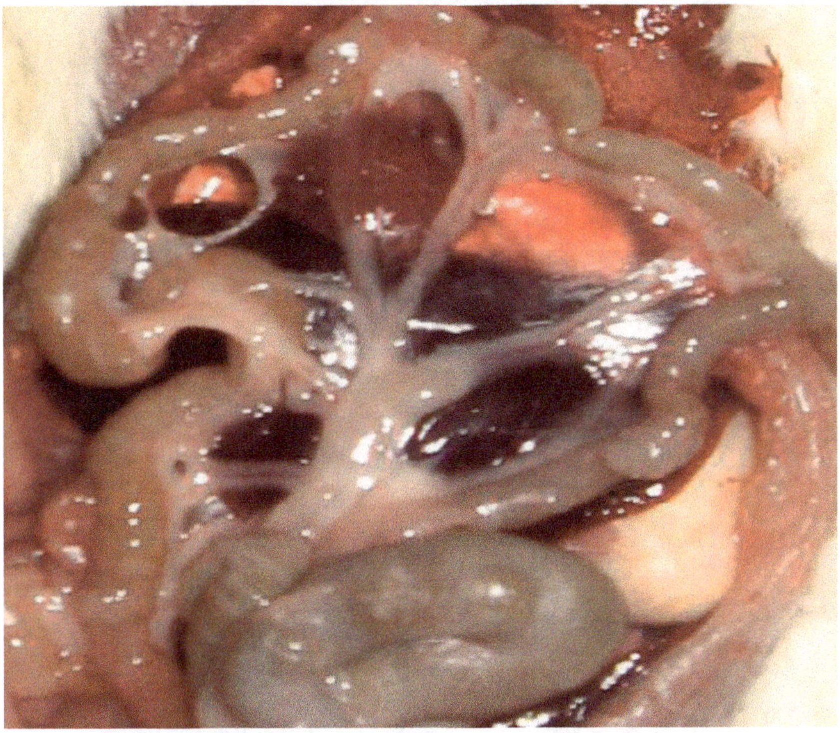

Figure 9. The intestine of rats in the first experimental group

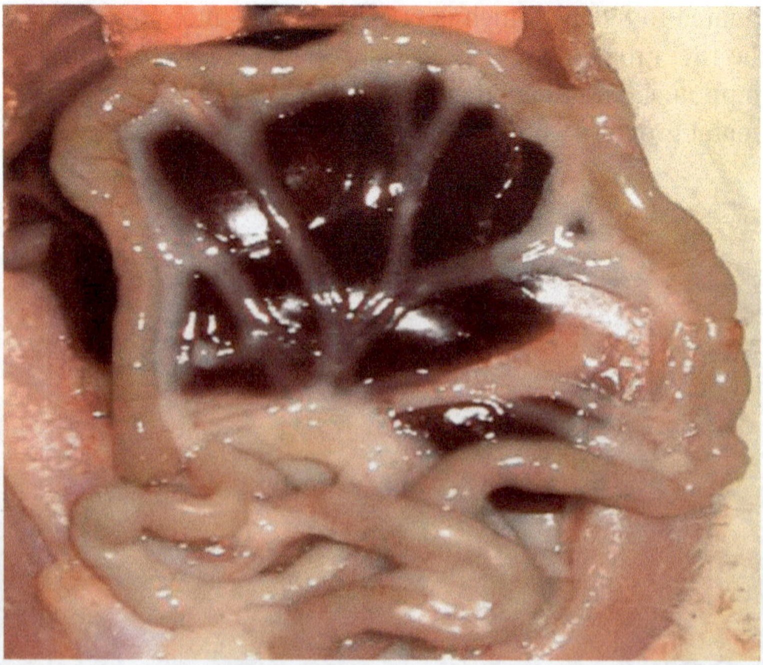

Figure 10. The intestine of rats in the second experimental group

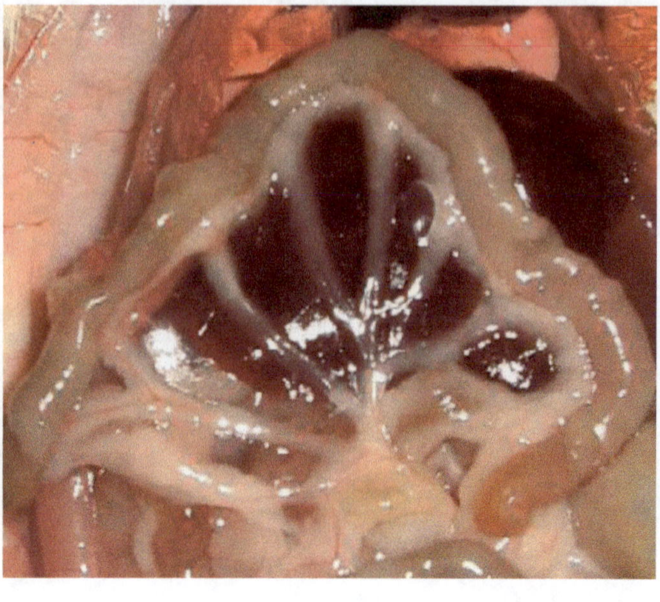

Figure 11. The intestine of rats in the third experimental group

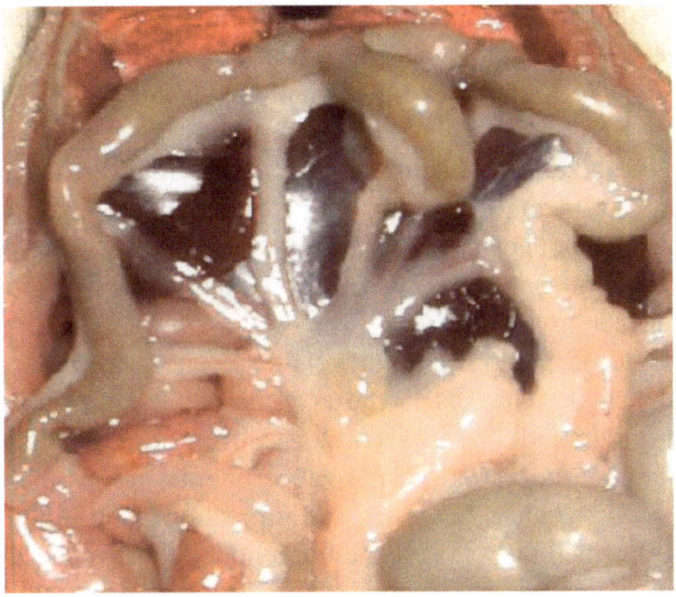

Table 17. Average width of mesenteric fat (Personal research data)

Group	Width of mesenteric fat	
	mm	%
Control	3.0 ± 0.1	100.0
1 experimental	3.0 ± 0.2	100.0
2 experimental	4.3 ± 0.1	143.3
3 experimental	6.2 ± 0.3	206.6

It was found out that in rats of the control group there were changes characteristic of serous and serous-catarrhal enteritis in the small intestine. On the mesentery, a small amount of body fat of a soft consistency was noted. The average width of mesenteric fat was 3 mm.

There were no any pathological processes in the intestine of animals of the first experimental group (kissel according to the recipe No. 590). Adipose tissue had a denser consistency than in rats of the control group, which indicates a favorable effect of kissel on the digestive canal. The average width of adipose tissue on the mesentery was 3 mm.

Focal serous enteritis was observed in rats of the second experimental group (briquetted concentrate, Russia). It should be noted that the adipose tissue on the mesentery had a denser consistency, its average width was 4.3 mm, which is 43% more compared to animals in the control and the first experimental groups.

Changes characteristic of serous-catarrhal enteritis were detected in rats of the third experimental group (powder concentrate, Poland). At the same time, significant fat deposits were noted on the mesentery, the average width of which was 6.2 mm, which is 106.6% more than in animals of the first and second experimental groups.

The results of morphological studies of the spleen and liver of experimental rats showed that the changes in the animals of the control and the first experimental groups did not significantly differ from the physiological norm. In the spleen of rats of the second and third experimental groups, micronecrosis, edema of the stroma connective tissue and rarefaction of the follicle substance were detected, which may indicate an immunodeficiency state.

The liver of rats of the second and third experimental groups has granular and diffuse fatty degeneration (up to the cricoid form). Diapedetic hemorrhages (especially in the third experimental group), discompletion of the beam structure, and impaired tinctorial properties of hepatocytes were revealed in these groups.Such changes indicate structural and functional disorders of the organ, and the depletion of the liver with glycogen, due to impaired carbohydrate metabolism with regular consumption of kissels from concentrates.

The results of morphological studies are presented in Figures 12.-15. *(Personal research data)*
The results of our studies showed the following:

- the quality level of kissels from concentrates is lower than that of kissel according to the traditional technological scheme, the nutritional and energy value vary significantly;
- not one kissel from concentrates contains the amount of vitamin C than that of kissel according to the traditional technological scheme.

In the study of oatmeal kissel, it was noted the following:

- according to the functional and technological properties of starch, the product "Oat breakfast" of the "Velle" trademark is the best. It has a mucous consistency and almost completely passed the stage of gelatinization of

Figure 12. Spleen of rats of the second experimental group: a) necrobiotic processes in the perifollicular zones. HE × 70; b) swelling of the stroma of the organ and the follicle substance. HE × 150

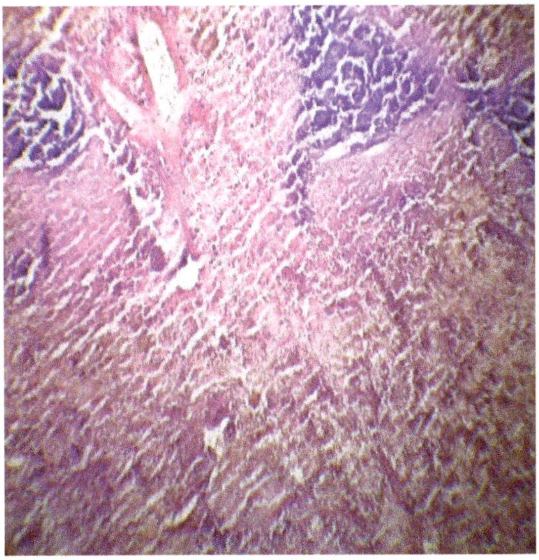

Figure 13. Liver of rats of the second experimental group: a) discompletion of the beam structure, perivascular and pericellular edema. HE × 70; b) diffuse fatty degeneration, violation of the tinctorial properties of hepatocytes. HE × 70

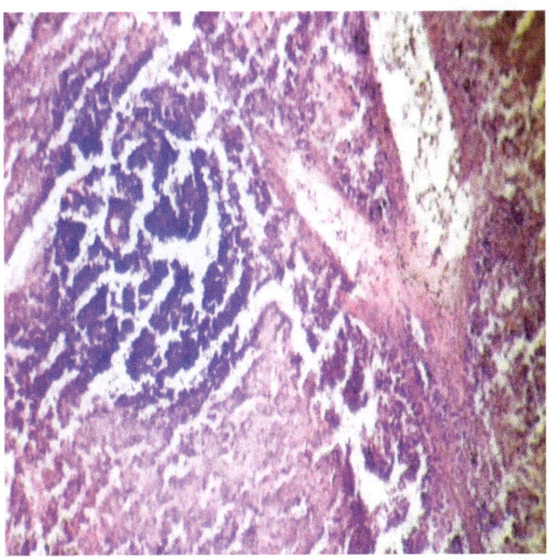

Figure 14. Spleen of rats of the third experimental group: a) micronecrosis in the red pulp. HE × 150' b) depression and swelling of the substance of the follicles. HE × 300

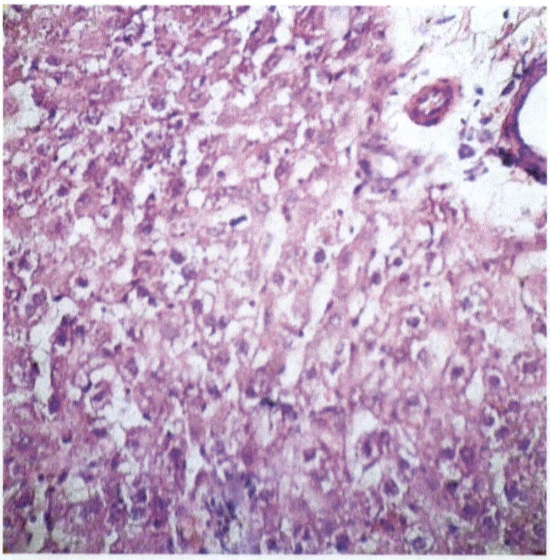

Figure 15. Liver of rats of the third experimental group: a) discompletion of the beam structure, granular and fatty degeneration. HE × 70; b) multiple diapedetic hemorrhages. HE × 70

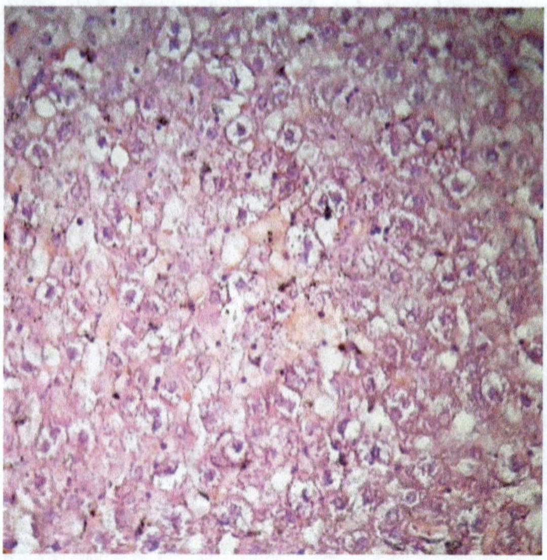

natural starch polysaccharides, it is able to exert a functional effect on the human gastrointestinal tract;

- kissel concentrate of the "Compass Zdorovya" brand, consisting of natural components, will not be able to exert a functional effect on the human gastrointestinal tract, since the method proposed by the manufacturer is insufficient for the full gelatinization of starch polysaccharides.

Consequently, to produce a functional product from natural ingredients, it is necessary to take into account technological factors that affect the maximum disclosure of the potential of the functional and technological properties of natural starch polysaccharides.

The research of the safety of kissels from concentrates with regular consumption revealed:

- regular consumption of kissels from concentrates contributed to the liveweight gain of experimental animals. In rats of the third experimental group, the weight deviation is 60.9% more than in animals of the control group;
- results of autopsy showed that rats fed with kissels from concentrates had significantly developed fat deposits on the mesentery (106.6% more than in the control group). Rats of the third experimental group had changes characteristic of catarrhal enteritis;
- rats fed with kissels from concentrates had granular and diffuse fatty degeneration (up to the cricoid form), discompletion of the beam structure and impaired tinctorial properties of hepatocytes, as well as diapedetic hemorrhages (especially in the third experimental group).

From the above it follows that regular consumption of kissel from concentrates can have an adverse effect on the body. This fact must be taken into account when developing diets, that is why the preference should be given to the desserts according to traditional technology. In addition, it is recommended to reduce the use of dessert concentrates in children's, dietary and therapeutic diet.

CONCLUSIONS OF THE CHAPTER

Information-analytical and own research of the authors allow us to state the following:

- analogues of traditional dishes (concentrates), which have firmly entered the life of a modern person, cannot fill the needs for the most necessary food components, such as proteins, fats and carbohydrates;
- replenishment of vitamins on the example of dessert (kissel) concentrates also cannot be guaranteed, despite the fact that some of the studied groups of concentrates were fortified. While kissel according to the traditional recipe allows one to get virtually fully predictable vitamins;
- fast-food products cannot provide nutritional value and energy demand. Refined components in the human body will be quickly absorbed, provoking the consumption of new portions of food;
- fats in the concentrates of soups and second courses (instant noodles, potato puree) are of a potential danger with the regular consumption of these products, due to transformation into substances (epoxides, insoluble copolymers in petroleum ether) during processing and storage. They reduce the body's resistance and provoke carcinogenic, mutagenic, and teratogenic effects;
- starches, which are the structure-forming component of dessert from concentrates, are not capable of exerting the expected therapeutic and prophylactic effect on the gastrointestinal tract, since they have not passed the full gelatinization stage;
- regular consumption of desserts from concentrates, unlike those according to the traditional recipe, can provoke a negative effect on the human body at the cellular level.

Thus, instant dishes (concentrates) cannot be a full substitute for dishes prepared according to traditional recipes and technologies. Regular consumption of concentrates and their prevalence in the diet poses a potential threat to the immune system, provokes the development of alimentary-dependent diseases, and in the future it can become a precursor of the development of carcinogenic, mutagenic, and teratogenic effects.

The production of concentrates that can not have a negative effect and (or) be really useful for human health requires a change in technological approaches on a global scale and new scientific and technical support that takes into account the culture of nutrition.

REFERENCES

Ackroff, K., & Sclafani, A. (2016). Flavor Preferences Conditioned by Dietary Glutamate. *Adv Nutr., 7*(4), 845-52.

American Diabetes Association. (2017). Classification and diagnosis of diabetes. *Diabetes Care, 40*, 11–24. doi:10.2337/dc17-S005

Atanazevich, V. I. (2000). *Drying food. Reference manual.* DeLi.

Blumenthal, M. M. (1991). A new look at the chemistry and physics of deep fat frying. *Food Technology, 45*(2), 68–74.

Bodinham, C. L., Smith, L., Thomas, E. L., Bell, J. D., Swann, J. R., Costabile, A., Russell-Jones, D., Umpleby, A. M., & Robertson, M. D. (2014). Efficacy of increased resistant starch consumption in human type 2 diabetes. *Endocrine Connections, 3*(2), 75–84. doi:10.1530/EC-14-0036 PMID:24671124

Bouchon, P., & Pyle, D. L. (2005a). Modelling oil absorption during post-frying cooling - I:Model development. *Food and Bioproducts Processing, 83*(4), 253–260. doi:10.1205/fbp.05115

Burich, O., & Berki, F. (1978). Drying of fruits and vegetables. Academic Press.

Carroll, K. K. (1989). Dietary fats and energy in relation to cancirogenesis. *New Era Global Harmony Nutr.: Proc. 14th Int. Congr. Nutr.*, 1, 493–496.

Chenlin, G., & Rao, M. (2019, November 15). Non-Digestible Carbohydrate and the Risk of Colorectal Neoplasia: A Systematic Review. *Nutrition and Cancer, 25*(Jun). Advance online publication. doi:10.1080/01635581.2020.1742360

Cho, N. H., Shaw, J. E., Karuranga, S., Huang, Y., da Rocha Fernandes, J. D., Ohlrogge, A. W., & Malanda, B. (2018). IDF diabetes atlas: Global estimates of diabetes prevalence for 2017 and projections for 2045. *Diabetes Research and Clinical Practice, 138*, 271–281. doi:10.1016/j.diabres.2018.02.023 PMID:29496507

Cohen, S.M., Eisenbrand, G., Fukushima, S., Gooderham, N.J., Guengerich, F.P., Hecht, S.S., Rietjens, C.M., Davidsen, J.M., Harman, C.L., & Taylor, S.V. (2018). Updated procedure for the safety evaluation of natural flavor complexes used as ingredients in food. *Food Chem Toxicol., 113*, 171-178. doi:10.1016/j.fct.2018.01.021

Dana, D., & Saguy, I. S. (2006). Review: Mechanism of oil uptake during deep-fat frying and the surfactant effect-theory and myth. *Advances in Colloid and Interface Science*, *128*, 267–272. doi:10.1016/j.cis.2006.11.013 PMID:17196541

Farrand, C., Charlton, K., Crino, M., Santos, J., Rodriguez-Fernandez, R., Ni Mhurchu, C., & Webster, J. (2017). Know Your Noodles! Assessing Variations in Sodium Content of Instant Noodles across Countries. *Nutrients, 9*(6). . doi:10.3390/nu9060612

Fukuzumi, K. (1978). Studies on the Reaction of Fats and Oils – Hydrolysis, Autoxidation, Hydrogenation, ets. *Mem. Fac. Eng. Nagoya Univ.*, *30*, 200–244.

Gau, E., Mate, D.M., Zou, Z., Oppermann, A., Töpel, A., Jakob, F., Wöll, D., Schwaneberg, U., & Pich, A. (2017). Sortase-Mediated Surface Functionalization of Stimuli-Responsive Microgels. *Biomacromolecules, 18*(9), 2789-2798. . doi:10.1021/acs.biomac.7b00720

Ginzburg, A. S. (1973). Fundamentals of the theory and technique of drying food products. Academic Press.

Glants, S. (1998). Biomedical statistics. Academic Press.

Jedrkiewicz, R., Kupska, M., Glowacz, A., Gromadzka, J., & Namiesnik, J. (2016). 3-MCPD: A worldwide problem of food chemistry. *Crit Rev Food Sci Nutr, 56*(14), 2268–2277.

Kodensova, V.M., Vrheshesinskaya, O. A., Risnik, D. V., Nikityuk, D. B., &Tutelyan, V. A. (2017). Provision of the population of Russia with micronutrients and the possibility of its correction. The state of the problem. *Nutrition, 86*(4), 113–124. . doi:10.24411/0042-8833-2017-00067

Makker, K., Alissa, R., Dudek, C., Travers, L., Smotherman, C., & Hudak, M.L. (2018). Glucose Gel in Infants at Risk for Transitional Neonatal Hypoglycemia. *Am J Perinatol., 35*(11), 1050-1056. . doi:10.1055/s-0038-1639338

Menyanu, E., Russell, J., & Charlton, K. (2019, June). Dietary Sources of Salt in Low- and Middle-Income Countries: A Systematic Literature Review. *International Journal of Environmental Research and Public Health, 16*(12), E2082. Advance online publication. doi:10.3390/ijerph16122082 PMID:31212868

Migunova, Y. V., & Sadykov, R. M. (2018). Nutrition of children in the modern Russian family: socio-economic aspect. *Nutrition, 87*(2), 103–107. . doi:10.24411/0042-8833-2018-10024

Novozhilova, G. P., Aksenova, V. M., & Mozgovaja, L. A. (2019). *The state of lipid peroxidation and antioxidant system in plasma, erythrocytes and saliva of children with diseases of the oral organs, aggravated by intestinal dysbiosis*. Available at: http:www.stomatburg.ru/articles/klin

Olijnik, J. V. (2007). Disorders of lipid peroxidation and their correction in children suffering from atopic dermatitis. *Bulletin of Scientific Research, 3*, 39–42.

Park, J., Lee, J.S., Jang, Y.A., Chung, H.R., & Kim, J. (2011). A comparison of food and nutrient intake between instant noodle consumers and non-instant noodle consumers in Korean adults. *Nutr Res Pract., 5*(5), 443-9. doi:10.4162/nrp.2011.5.5.443

Rzhehin, V. P., & Sergeev, A. G. (1967). *Determination of the total content of oxidation products insoluble in petroleum ether. "Guide to research methods, techno-chemical control and accounting of production in the oil and fat industry"*. All-Russian Research Institute of Fats Publ.

Sacilik, K., Keskin, R., & Elicin, A. K. (2006). Mathematical modelling of solar tunnel drying of thin layer organic tomato. *Journal of Food Engineering*, *73*(3), 231–238. doi:10.1016/j.jfoodeng.2005.01.025

Sajilata, M. G., Singhai, R. S., & Kulkarni, P. R. (2006). Resistant starch: A review. *Comprehensive Reviews in Food Science and Food Safety*, *5*(1), 1–17. doi:10.1111/j.1541-4337.2006.tb00076.x PMID:33412740

Sanders, T. (1987). Toxicological considerations in oxidative rancidity of animal fats. *Food Science & Technology Today*, *1*(3), 162–164.

Shen, L., Keenan, M. J., Raggio, A., Williams, C., & Martin, R. J. (2011). Dietary-resistant starch improves maternal glycemic control in Goto-Kakazaki rat. *Molecular Nutrition & Food Research*, *55*(10), 1499–1508. doi:10.1002/mnfr.201000605 PMID:21638778

Simakova, I. V. (2004). *Research of palm oil in the technology of frying products*. Saint-Petersburg.

Simakova, I. V. (2015). *Scientific and applied safety aspects of fast food products*. Orel.

Simakova, I.V., Eliseev, J.J., Perkel, R.L., Domnickij, I.J., Terent'ev, A.A., Strizhevskaja, V.N., & Makarova, A.N. (2016b). Evaluation of sanitary and technological safety of the fatty component of certain types of snacks and flour confectionery products with a long shelf life. *Saratov Scientific Medical Journal*, *12*(3), 333–339.

Simakova, I. V., Perkel, R. L., Zakrevkij, V. V., & Kutkina, M. N. (2016a). The study of the intensity of pathogenesis, depending on the degree of oxidation of palm oil. *Voprosypitanija*, *85*(2), 36.

Stopskij, V.S., Melamud, N.L., Kulichenko, G.E., & Jestrina, F.B. (1968). *The method of quantitative determination of epoxy groups in fats.* Author's certificate USSR, no 1040914, IP C7G 01 N 33/02, G 01N 31/02.

Vysheslavova, M. J. (1969). Effect of sunflower oil with different degrees of oxidation on the induction of 2-acetylaminofluorene tumors in rats. *Voprosy Onkologii*, *15*(4), 66–70.

Wójcik-Stopczyńska, B., Falkowski, J., & Jakubowska, B. (2002). Microbiologic evaluation of instant soup concentrates. *Roczniki Panstwowego Zakladu Higieny*, *53*(2), 149–156. PMID:12235671

Zhou, J., Martin, R. J., Tulley, R. T., Raggio, A. M., McCutcheon, K. L., Shen, L., Danna, S. C., Tripathy, S., Hegsted, M., & Keenan, M. J. (2008). Dietary resistant starch upregulates total GLP-1 and PYY in a sustained day-long manner through fermentation in rodents. *American Journal of Physiology. Endocrinology and Metabolism*, *295*(5), 1160–1166. doi:10.1152/ajpendo.90637.2008 PMID:18796545

Conclusion

Globalization, industrialization, and urbanization have already influenced the authenticity of human nutrition and increased the risks of a decrease in the body's resistance due to the lack of essential substances. Regular consumption of fast-food products leads to a monotonous supply of nutrients that are significantly different from natural ones. Global foodstuffs are depleted in essential substances that can support the human immune system. The COVID-19 pandemic has shown that people who have antioxidants in their diet recover faster and have a higher body resistance.

According to numerous studies, the authors of the book declare that nutrition is a critical factor in the direct impact on human health in the long term. The growth of nutritionally dependent diseases in children and adults is becoming threatening. In adulthood, childhood diseases and pathologies progress and significantly worsen the quality of life. It is important to conduct research on the effects of modern foods on human health in the long term (with regular consumption).

From the data presented by the authors, it is obvious that the consumption of fast food in the world is of a global nature. The network of fast-food enterprises is constantly developing and modernizing, and the popularity of the products does not decrease. This fact should be taken into account in predicting the health of the population regularly consuming fast food products. In modern realities, a record of the adequate consumption of essential nutrients, especially irreplaceable ones, is becoming an acute problem for mankind.

The authors presented their own classification according to combined characteristics, which most fully reflects the entire range of fast-food products manufactured in public catering and industrial enterprises.

The presented analysis of the prevalence of global catering enterprises in the world and the analysis of technological solutions used in these enterprises allows one to single out several unifying aspects for fast food products:

- the main ingredients that unite most fast-food industries are fats and starch contained in various types of flour (wheat, corn, rice);
- the main technological methods for preparing products before selling or storing are frying or processing under pressure and high temperatures.

The authors have established and confirmed by the data of the world scientists that the processes of thermal oxidation of frying fat at temperatures of 170-180 °C differ from the standard fats autooxidation at moderate temperatures due to the increased oxidative activity of the peroxyl radical LOO• and LOOH hydroperoxides.

As a result of oxidation in deep-frying fats, epoxies and dioxides, insoluble in petroleum ether (PETRO), accumulate. A correlation has been established between the content of epoxides and RRPE.

Peroxides at temperatures of 170-180 °C are intermediate oxidation products and cannot be used to control the oxidation state of frying fat. Free fatty acids are formed only as a result of the hydrolysis of frying fat. The accumulation of free fatty acids correlates with the content of polar compounds determined by the electrophysical method. However, neither the determination of the acid number with test strips nor the determination of polar products can be used to determine the oxidation state of frying fat.

Analysis of the results of a series of experiments conducted on more than 200 experimental animals (white rats) with the study of various internal organs: liver, large and small intestines, stomach, kidneys and myocardium, with a high degree of reliability showed the negative physiological effect of fat oxidation products on the body. All changes in the groups of rats that consumed global food products with an oxidized fat phase and fats used for frying them were a manifestation of inhibition of hematopoietic activity, hemodynamic and dystrophic disorders, an immunodeficiency state and inflammatory processes of varying severity.

Such diverse experiments have been carried out for the first time. Studies, extended by hematological data, revealed that fat oxidation products cause serious changes in the biochemical composition and blood cells, and lead to disruptions in the work of the body's antioxidant defense systems.

Experiments on animals have proved a close relationship between the content of ETCs and their effect on the body and the need for regulation and standardization of this indicator in finished products at the stages of their circulation and sale to ensure the safety of fast food products.

Information-analytical and authors' own research suggests that analogs of traditional dishes (concentrates) that have firmly entered the life of a modern person cannot meet the needs for the most essential food components, such as proteins, fats and carbohydrates. Products intended for quick consumption cannot provide not only nutritional value, but also energy requirements. Refined components in the human body will be quickly absorbed, provoking the consumption of new portions of food.

Products containing starches, which are a structure-forming component, are not able to provide the expected therapeutic and prophylactic effect on the gastrointestinal tract, since they do not go through the full stage of gelatinization when quickly cooked.

Thus, instant food and concentrates cannot be a complete substitute for dishes prepared according to traditional recipes and technologies. Regular consumption of concentrates and their prevalence in the diet poses a potential threat to immunity, provokes the appearance of nutritionally dependent diseases, and in the future can become a precursor to the occurrence of carcinogenic, mutagenic, and teratogenic effects.

Obviously, global nutrition technologies require the development of special approaches that take into account the preservation of essential components and contribute to the preservation of the generation health. Such issues can be solved by uniting specialists who develop food products of a new generation, taking into account the transformation of nutrients in the technological process and storage process, as well as the metabolism.

Trends in the development of specialized food products, as well as products and diets of personalized nutrition (based on taking into account the allelic polymorphism of individual "predisposition genes" associated with the assimilation of certain food nutrients) are relevant today, aimed at improving the nutrition structure of the population as a whole and individual groups. When formulating personalized individual diets, consider the following:

- the need for nutrients, depending on the person state;
- the body's ability to assimilate nutrients, that is, the enzymatic and microbiotic activity of the body, in combination with the state of the digestive system as a whole, as well as the immune system;
- individual preferences with adequate product interchangeability.

This book calls on scientists and technologists who develop global food to change the way they approach modern technological solutions. It is also necessary to pay attention to the need for enhanced quality control of fatty components that make up global food products and fats used as a technological factor (deep fat).

About the Authors

Inna Vladimirovna Simakova was born on 20 November 1977 in Saratov city, Russia. She is married and has two children. Inna graduated from Saratov State Agrarian University in 2000 and was qualified as an engineer in the specialty of Technology of public catering products. From 2000 to 2004 she has been studying in State Trade & Economy University, Department of food science, St.Petersburg Russia and graduated as a Ph.D. During couple years (from 2013 to 2015) she got a Doctor degree of Technical Sciences (Department of food science, Prioksky State University, Russia). From graduating to now she has been working in Saratov State Agrarian University named after N.I.V avilov, Russia. Her professional activities include her working as a director of the small innovative enterprise "Healthy Nutrition" (2012-2022). In September 2021, textbooks in Innas co-authorship will be published for the areas of training "Food Sciences" of bachelor's, master's and postgraduate studies: "Technology of public catering products" and "Scientific and practical aspects of technology of food industry products". She has 15-years of experience in the development of diets for the social sphere, including therapeutic, dietary and preventive diets (for children with phenylketonuria, cystic fibrosis, Down's disease, celiac disease, obesity, diabetes mellitus, microelementosis): kindergartens, schools, children's health camps. Also she has experience in the technological design of food industry enterprises with the layout of premises and equipment, taking into account the technological process and organization of production and plumbing calculations of the enterprise's need for electricity, water, etc. Most of scientific research is devoted to the study of the safety or effectiveness of food products in experiments in vivo. Over the past few years, the focus of her scientific activity has been directed to the interdisciplinary areas of preventive medicine. She conducts research together with medical scientific community of Saratov State Medical University named after V.I. Razumovsky. With the permission of the local ethics committee, together with doctors, she holds clinical studies of the effectiveness of the products of therapeutic, dietary and preventive and functional nutrition developed by her scientific school on patients with metabolic syndrome.

Victoriya Nikolaevna Strizhevskaya works at the Department of Food Technology from 1997 to the present. She graduated from the Saratov State Academy of Veterinary Medicine and Biotechnology with a degree in "Technology of meat and meat products", qualification - "Engineer-technologist". Theme of the thesis: "The study of technological properties and ways of using lamb in public catering" defense took place on January 28, 2004 in the dissertation council of the St. Petersburg Trade and Economic Institute. Direction of research activities: 05.18.15- "Technology and commodity research of products of functional and specialized purposes and catering." He has more than 40 publications, 30 publications in the RSCI, 5 publications in journals according to the list of HAC, 8 teaching aids for bachelors and masters in the areas of training on "Technology products and the organization of public catering".

Roman Lvovich Perkel, born in 1938, graduated from the Leningrad Technological Institute of the Refrigeration Industry in 1961, the Department of General Technology and Refrigeration Technology of Food Products, majoring in Preservation Technology, qualified as a Process Engineer in the Refrigeration Industry. In 1970, Roman L. Perkel defended his thesis for the degree of Candidate of Technical Sciences, "The study of autolytic changes in tissue fat of frozen herring during refrigerated storage." From 1969 to 2001, Roman L. Perkel worked at the All-Russian Scientific Research Institute of Oil Industry of the Russian Academy of Agricultural Sciences, St. Petersburg, consistently as a junior researcher, senior researcher, senior researcher, Ph.D., and from 1972 to 2001 as a Head of Laboratory of Transesterification of Fats. In 1977, by the decision of the Higher Attestation Commission, Roman L. Perkel was awarded the academic title of Senior Researcher majoring in Technology of Fats, Essential Oils and Perfumery and Cosmetic Products (no. 05.18.06). In 1990, Roman L. Perkel defended his thesis in the form of a scientific report for the degree of Doctor of Technical Sciences, "Research and development of technology and organization of interesterified edible fats production". In 1991, by the decision of the Higher Attestation Commission, Roman L. Perkel was awarded the degree of Doctor of Technical Sciences. Since 1993, Roman L. Perkel has been a Corresponding Member of St. Petersburg Engineering Academy. From 2001 to 2004, Roman L. Perkel worked in the position of Deputy General Director for Science, Salolin JSC, St. Petersburg. From 2004 to 2016, Roman L. Perkel worked as Professor of the Department of Technology and Catering, St. Petersburg State University of Trade and Economics. In 2013, by the decision of State Commission for Academic Degrees and Titles, Roman L. Perkel was awarded the title of Professor in the Department of Technology and Catering. From September 2016 to June 2017, Roman L. Perkel worked as a professor at the Graduate School of Biotechnology and Food Technologies, St. Petersburg Polytechnic University of Peter the Great.

Galina Yuryevna Rakhmanova was born in 1963 in Kalinin, Kalinin region. She is married and has two children. Galina graduated from Saratov State Medical University in 1986 and was qualified as a pediatrician. After graduating from university, she has been working in sanitary and epidemiological organizations as epidemiologist as the head of environmental sanitation department and the head of sanitary and epidemiological department. Nowadays, she is the head of children hygienic supervision department in the Rospotrebnadzor (Federal Service for Supervision of Consumers Protection) in Saratov region. Her experience as epidemiologist is over 32 years. Galina has a certificate as children`s hygienist. Professional obligations include the ensuring of the children's sanitary and epidemiological health in Saratov region by organizing the work on creating and functioning of the control system, monitoring sanitary and epidemiological state of objects of mass concentration of children. The department organizes the prevention of infectious diseases caused by pathogens and parasites, non-infectious diseases (intoxication) of children caused by xenobiotics and superecotoxicants. The aim of her work is to identify the reasons and conditions of occurrence and spreading of infection and parasitic diseases, and also massive non-infectious ones (intoxication) on the objects for children by special sanitary and epidemical investigations. She defines the casual nexus between children's health and their inhabitancy. Besides, she carries out sanitary-epidemiological examination, investigations of the spots of children location. Moreover, she widely promotes the healthy lifestyle, healthy food to support the balance of activity and rest, and also she does the best to prevent bad habits among children.

Index

www.ingramcontent.com/pod-product-compliance
Lightning Source LLC
LaVergne TN
LVHW081520300926
849161LV00006B/1232

* 9 7 8 1 7 9 9 8 8 1 9 7 1 *